PRE-GED

Math Workbook

REVIEW, PRACTICE AND METHOD

LEARN EVERYTHING YOU NEED IN 12 HOURS

Inside this book you will find

* 1,000+ practice questions for all topics

* 7 Full Revisions Guide for each topics

* Full length practice with time-line

Edition 2017

Ankur Sharma

About Author

Ankur Sharma is an experienced professor who had been teaching math and science for more than fifteen years at AIMS located in Thailand. He earned a B.Engineering a M.SC in Technology Management from Assumption College with full scholarship. Most of the students who studied GED Math and GED Science have pass the exam easily with a score of 500 above. He is also specialized in teaching SAT Physics and Math level 1 and level 2. His goal for writing this book was to help students and examinee to use the time efficiently thru the right material with a proper method.

Feel free to ask any question on facebook page: facebook.com/IGEN4U

ISBN-13: 978-1546531128
ISBN-10: 1546531122
BISAC: Education / Adult & Continuing Education

This book will help you prepare for GED Math Examination. Contents inside are up to date, there are plenty of practice exercises and examples. GED Mathematical Reasoning Test focuses on two major content areas: quantitative problem solving and algebraic problem solving. GED Mathematic exam assess important mathematical proficiencies, including modeling, constructing and critiquing reasoning, and procedural fluency.

What you should know about GED Mathematic :

- About 45 percent of the content in the test focuses on quantitative problem solving, and 55 percent focuses on algebraic problem solving.
- You will be tested on procedural skill and fluency as well as problem solving.
- The contexts within which problem solving skills are measured were taken from both academic and workforce contexts.
- About 50 percent of the items are written to a depth of knowledge cognitive complexity.
- About 30 percent of the items are aligned to a mathematical practice standard in addition to a content indicator.
- The passing score for each subject is 410 out of 800

The exams are designed to measure the skills and knowledge that a student must acquired after four years in high school.

Exam structure are as follow:

	Number of Question	Time	Content
Part I (No Calculator)	5 questions	10 minutes	Number and Operations 20 - 30% Geometry 20 - 30%
Part II (Calculator allowed)	40 questions	105 minutes	Data analysis and Probability 20 - 30% Algebra and Functions 20 - 30%

Content

What's inside this book ?

In this book you will find questions that will improve your basic, a revision to check your understanding and a time-line that will help you pace up during the test.

How to prepare for GED-Maths examination ?

step1: take the pretest and see your score

step2: go through the work-book

step3: do the post-test and see the improvement

How does the scoring works for GED (2016) ?

New GED have change a full score of 800 to 200, this means new standard of score conversion too.

To understand scoring range we can summarize them into categories below:

- 145 to 164 → Pass/ High School Equivalency
- 165 to 174 → GED College Ready
- 175 to 200 → GED College Ready + Credit

Tips and Tricks

1. Read the question carefully and identify the main point

2. Understand the question and list out required operation

3. Pace yourself, note that time is limited

4. Flag the questions that are hard for you and come back to it later

5. Do not leave any questions blank (try to make an educated guess)

6. Make use of the calculator and formulas given

7. Eliminate choices

Pre-Test

(40 minutes)

Part 1 (calculator allowed)

1. What is the volume (in centimeter cube) of a sphere with radius 9 cm.?

a) 81π
b) 108π
c) 256π
d) 728π
e) 972π

2. What is the sum of the interior angle of hexagon (6-side polygon)?

a) 720°
b) 640°
c) 540°
d) 480°
e) 360°

3. Evaluate 2 + 5 ÷ 0.1 x 2 -3?

a) 122
b) 99
c) 67
d) 34
e) 4

4. Five oranges and two apples cost total of $ 5.00. Two oranges and two apples cost $ 4.00.   What is the cost of three oranges?

a) 1
b) 1.5
c) 2
d) 2.5
e) 3

5. If a = 1, b = -2, c =-1 What is the value of $3b^2 - 2a + c$?

a) 3
b) 4
c) 8
d) 9
e) 11

6. If x is a negative integer, which of the following gives the highest value?

a) $\dfrac{x}{01}$
b) $\dfrac{x}{001}$
c) $\dfrac{x}{10}$
d) $\dfrac{x}{100}$
e) $\dfrac{x}{1000}$

7. What is the shortest distance between point 'a' and 'b' if the radius of the circle is 4 units?

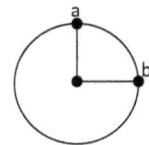

a) $\sqrt{2}$
b) $3\sqrt{2}$
c) 4
d) $4\sqrt{2}$
e) $5\sqrt{2}$

8. Which of the following line is perpendicular to line y = 10x -12

a) y = -0.1x + 12
b) y = -10x -12
c) y = x + 10
d) y = 0.1x -10
e) y = 10x -12

9. A cubic dice is thrown twice what is the probability that the sum of the faces will be a square number?

a) 3/6
b) 5/6
c) 3/36
d) 5/36
e) 9/36

10. Which expression gives the area of the figure below?

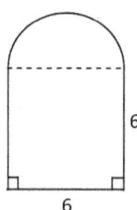

6

6

a) 6(3) + π (3)
b) 6(3) + π (4.5)
c) 6(6) + π (2)
d) 6(6) + π (3)
e) 6(6) + π (4.5)

Part 2 (non-calculator)

11. What is the probability of picking up red socks from a bag containing 5 blue, 3 green and 2 red socks ?

a) 1%
b) 2%
c) 10%
d) 20%
e) 50%

12. Find the value of 'a' if 5 : a = 30 : 126

a) 3
b) 6
c) 12
d) 18
e) 21

13. What is the slope of the line below ?

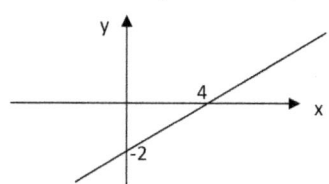

a)   0

b) $-\dfrac{1}{2}$

c) $\dfrac{1}{2}$

d) -2
e)   2

14. Jason deposited $ 40,000 in a saving account for 3 years which earn 5% simple interest annually. By the end of three year how much money will Jason have ?

a) 3,000
b) 6,000
c) 12,000
d) 43,000
e) 46,000

15. Find the missing angle 'w'

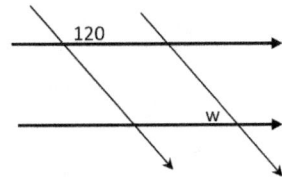

a) 30
b) 60
c) 90
d) 100
e) 120

16. Ian wants to pave a field with rectangular concrete solid blocks, each of a dimension of 36 inches by 36 inches by 6 inches. If a field requires 3 cubic yard of concrete then how many blocks are required ? (36 inches = 1 yd)

a) 18
b) 36
c) 60
d) 82
e) 120

17. Allan joined a charity ride to help cancer patients. On Monday he rode 36 miles, on Tuesday he rode 42 miles, on Wednesday he rode 28 miles, on Thursday he rode 33 miles and on Friday he rode 51 miles. What is the mean number of miles ridden by Allan ?

a) 32
b) 38
c) 40
d) 42
e) 46

18. 5, 11, 23, __ , 95
What is the value is the missing term of the sequence ?

a) 29
b) 42
c) 46
d) 47
e) 59

19. Where does the square of root of 92 lies ?

a) between 6 and 7
b) between 7 and 8
c) between 8 and 9
d) between 9 and 10
e) between 10 and 11

20. Which of the following is a solution of x^2 + 3x -28 = 0 ?

a) -7 only
b) 4 only
c) - 4 and 7
d) -7 and 4
e) 4 and 7

Answer Key

Part 1

1. e

2. a

3. b

4. a

5. d

6. c

7. d

8. a

9. d

10. e

Part 2

11. d

12. e

13. c

14. e

15. b

16. a

17. b

18. d

19. d

20. d

Scoring Range

Correct Number	Scoring Range
1 - 5	100 - 120
6 - 8	120 - 140
9 - 12	140 - 160
13 - 16	160 - 170
17 - 18	170 - 180
19 - 20	180 - 200

Time-Line in Pacing Yourself

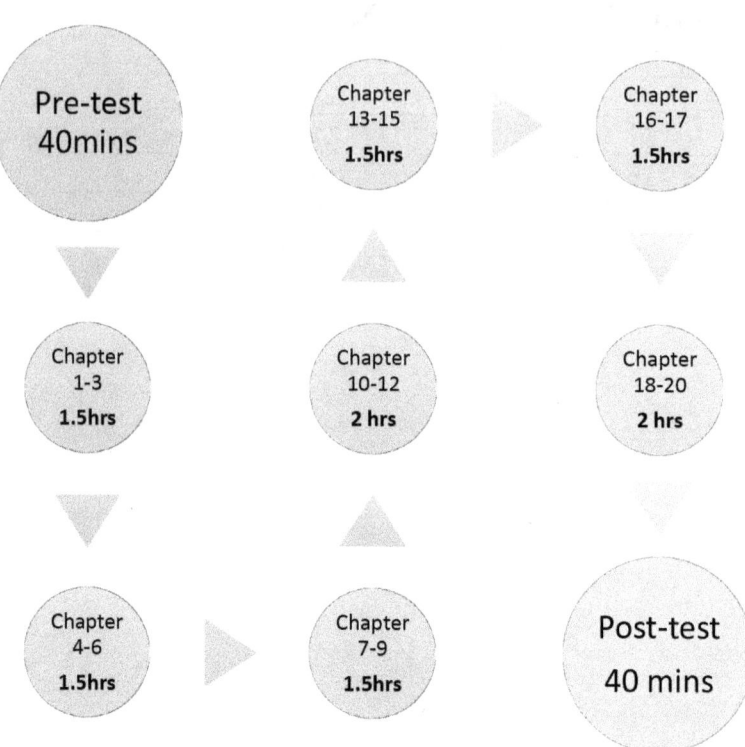

Chapter 1

Number Facts

1. Integers   are number that contain no decimals point.

	Examples
Positive integer	1, 2, 3, 4, 5
Negative integer	-4, -3, -2, -1
Zero (is neither positive nor negative)	0
Whole number	0,1,2,3,4
Even integer	-4, -2, 0, 2, 4
Odd integer	-3, -1, 1, 3
Consecutive Integers	1,2,3,4,5...44,45,46,47...91,92,93...
Consecutive Odd Integers	-3,-1,1,3,5...45,47,49....91,93,95....
Consecutive Even Integers	-4,-2,0,2,4....44,46,48....92,94,96...

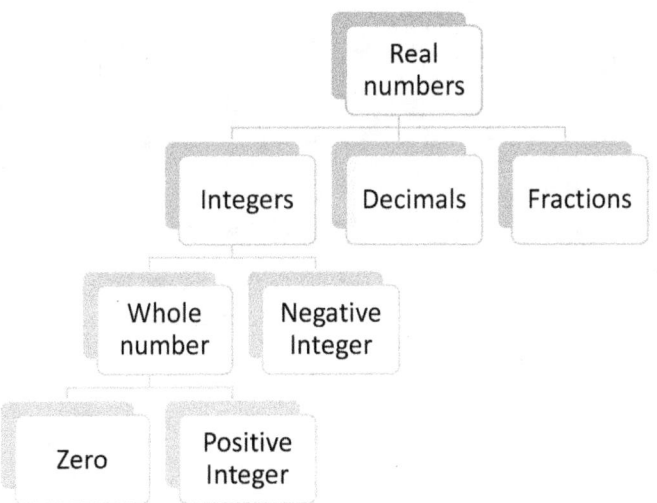

2.1 Non-Integers   are number that are either fractions or **decimals**

	Example:
Decimals can be **positive**	0.14, 1.2685, 22.349
Decimals can be **negative**	-0.56, -2.347 , -47.33

2.2 *Rounding decimals* if the number *behind the rounded digit is 5,6,7,8 or 9 then round it up by +1*

Round each number to the nearest whole number

i) 3.45819 → just look at 3.4 → 3 + 0 = 3

ii) 0.624 → just look at 0.6 → 0 + 1 = 1

iii) 16.7892 → just look at 16.7 → 16 + 1 = 17

Round each number to 3 significant figures

i) 3.45819 → just look at 3.458 →3.45 + 0.01 = 3.46

ii) 0.624 → just look at 0.6240 → 0.624 + 0 = 0.624

iii) 16.7892 → just look at 16.78 → 16.7 +0. 1 = 16.8

Round each number to 2 decimal place

i) 3.45819 → just look at 3.458 →3.45 + 0.01 = 3.46

ii) 0.624 → just look at 0.624 → 0.62 + 0 = 0.62

iii) 16.7892 → just look at 16.789 → 16.78 +0.0 1 = 16.79

3. Number values are measure and read according to the digits

billions	hundred-millions	ten-millions	millions	hundred-thousands	ten-thousands	thousands	hundreds	tens	units
__	__	__	__	__	__	__	__	__	__

Read the following numbers:

34,500 → thirty four ten-thousand five hundred

3,000,500 → three million five hundred

Round each number to the nearest ten

i) 234 → just look at 234 →230 + 0 = 230

ii) 5956 → just look at 5956→ 5950 + 10 = 5960

iii) 98 → just look at 98 → 90 +10 = 100

Round each number to the nearest thousand

i) 3432 → just look at 3432 →3000 + 0 = 3000

ii) 5756 → just look at 5756→ 5000 + 1000 = 6000

Round each number to 3 significant figures

i) 345819 → just look at 345819 →345000 + 1000
= 346000

ii) 5756 → just look at 5756 → 5750 + 10 = 5760

Exercise 1

1. Circle even numbers: 5 , 50 , -32 , 23 , -125 , 124 , 5678 , 221, 679 , 0

2. Circle positive even integers: 5 , 50 , -32 , 23 , -125 , 124 , 5678 , 221, 679 , 0

3. List the four whole number after: 0 , 1 , 2 , __ , __ , __ , __

4. List five more consecutive even integers after 32:   32 , __ , __ , __ , __ , __

5. List five more consecutive odd integers after -11:   -11 , __ , __ , __ , __ , __

6. List five more consecutive integers after -10:   -10 , __ , __ , __ , __ , __

7. List five more consecutive integers after 10:   10 , __ , __ , __ , __ , __

8. Fill in the boxes

Numbers	Round to nearest Integer	Round to nearest hundred	Round to 3 significant figures
56935.67			
234.19			
6003			
472274.72274			

9. Fill in the boxes

Numbers	Round to nearest integer	Round to 2 decimal place	Round to 3 significant figures
2.1956			
63.412			
6.006			
74.9999			

Chapter 2
Basic Operations

Addition or Sum→ **+**

13,458 + 2959 = ?
Placing integers in the aligned digit order first then add them

```
        13,458
    +    2,959
        16,417
```

9.0026 + 0.127 = ?
Placing integers in the aligned digit order and decimal place first then add them

```
        9.0026
    +   0.1270
        9.1296
```

Subtraction or **Difference** or **Minus→** **-**

13,458 - 2959 = ?
Placing integers in the aligned digit order first then minus them

```
        13,458
    -    2,959
        10,499
```

9.0026 - 0.127 = ?
Placing integers in the aligned digit order and decimal place first then minus them

```
        9.0026
    -   0.1270
        8.8756
```

Multiplication or **Product →** **X**

2,105 X 32 = ?
Placing integers in the aligned digit order first then multiply

```
           2,105
    X         32
           4210      ← (2105 x 2)
           6315□     ← (2105 x 3)
          67,360     ← ( add two lines
                         above)
```

2,105 X 3,200 = ?
Placing integers in the aligned digit order except the zeros then multiply

```
           2,105
    X       3200
             00
          4210□□    ←(2105 x 2)
          6315□□□   ← (2105 x 3)
       6,736,000    ← ( add three
                         lines above)
```

$4532 \div 2 = ?$

$550 \div 22 = ?$

$$
\begin{array}{r}
2266 \\
2\overline{)4532} \\
-4 \\
\hline
05 \\
-4 \\
\hline
13 \\
-12 \\
\hline
12 \\
-12 \\
\hline
0
\end{array}
$$

$22\overline{)550}$

B	O	D	M	A	S
Brackets	**Operations**	**Division**	**Multiplication**	**Addition**	**Subtraction**
(), {} , []	√, ^	÷	X	+	-

Example 1
What is the answer to this question ?

$40 - 3 \times 10 = ?$

a) 10
b) 370

Example 2
What is the answer to this question ?

$(40 - 3) \times 10 = ?$

a) 10
b) 370

Exercise 2

1. Evaluate the following sum

     a) 23,679 + 112 =

     b) 561,234 + 31,000 =

     c) 79.2 + 0.312 =

     d) 0.945 + 0.00405 =

2. Evaluate the following differences

     a) 32,550 - 2134 =

     b) 1,260,450 - 45,000 =

     c) 15.34 - 10.0019 =

     d) 0.0219 - 0.0086 =

3. Evaluate the following product

     a) 342 x 22 =

     b) 692 x 3000 =

     c) 1112 x 999 =

4. Evaluate the following quotient

     a) 5400 ÷ 60 =

     b) 3255 ÷ 5 =

     c) 20,104 ÷ 25 =

5. Evaluate the following

     a) 50 + 50 ÷5 =

     b) 323 x 15 – 3 =

     c) (6000 - 300) x 25 =

Chapter 3
Special Numbers

Factors   of a number are positive integers that can be divided into the number without the remainder.

Example:

Factors of 2 → 1, 2

Factors of 12 → 1, 2, 3, 4, 6, 12

Factors of 36 → 1, 2, 3, 4, 6, 9, 12, 18, 36

Prime number   is a number that has only 2 factors (itself and 1).

Example: 2, 3, 5, 7, 11, 17, 19, 23, ...

Multiples of a number are integers that are the results of the number multiplied by any integers.

Example:

Multiple of 3 → 3, 6, 9 ,12, 15, 18, ...

Multiple of 10 → 10, 20, 30, 40, 50, ...

Multiple of 36 → 36, 72, 108, 144, ...

Square number   is a number raise to power of 2 (the number times itself once).

Example:

$1^2 = 1 \times 1 = 1$

$2^2 = 2 \times 2 = 4$

$3^2 = 3 \times 3 = 9$

$4^2 = 4 \times 4 = 16$

$5^2 = 5 \times 5 = 25$

$6^2 = 6 \times 6 = 36$

$7^2 = 7 \times 7 = 49$

$8^2 = 8 \times 8 = 64$

$9^2 = 9 \times 9 = 81$

$10^2 = 10 \times 10 = 100$

Cube number is a number raise to power of 3 (the number times itself twice).

Example:

$$1^3 = 1 \times 1 \times 1 = 1$$
$$2^3 = 2 \times 2 \times 2 = 8$$
$$3^3 = 3 \times 3 \times 3 = 27$$

$$4^3 = 4 \times 4 \times 4 = 64$$
$$5^3 = 5 \times 5 \times 5 = 125$$
$$6^3 = 6 \times 6 \times 6 = 216$$

Remainder is what's left over when two number cannot be divided perfectly.

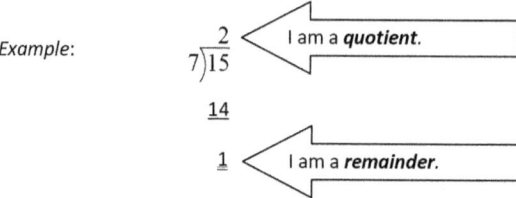

Example:

Mean or *average*   is the sum of all values divided by the number of terms.

Mode   is the most occurred value.

Median   is the middle number after the list is arranged in order of size.

Example:

A = {1, 3, 3, 3, 4, 5, 8, 8, 10}

Set A is shown above, find mean (arithmetic mean), mode, and median.

Solution:

$$\text{Mean (Arithmetic Mean)} = \frac{\text{Sum of values}}{\text{Number of values}}$$

$$= \frac{1 + 3 + 3 + 3 + 4 + 5 + 8 + 8 + 10}{9} = \frac{45}{9} = 5$$

Mode = {1, 3, 3, 3, 4, 5, 8, 8, 10} = 3

Median = {1, 3, 3, 3, 4, 5, 8, 8, 10} = 4

Exercise 3

1. List three prime numbers after

     a) 23 →

     b) 53 →

     c) 101→

2. List all the factors of the following :

     a) 30 →

     b) 45 →

     c) 100 →

3. List all the factors of 10a →

4. List all the prime factors of 69 →

5. List all the prime factors of 100 →

6. Evaluate the following :

$13^2 =$ $100^2 =$ $9^3 =$

$15^2 =$ $700^2 =$ $10^3 =$

$20^2 =$ $999^2 =$ $20^3 =$

7. What is the remainder :

     a) When 25 is divided by 7?

     b) When 99 is divided by 6?

8. What one digit number divided by 5 gives the remainder of 4?

9. Anna weighs 45 kg, Sui weighs 52 kg, Marc weighs 61 kg and Jacob weighs 43 kg.

     a) What is the mean and median of their weight?

     b) If Louis weighs 52 kg joined the group, what would be the mean, median and mode of their weight?

10. If the average of x, 10 and 15 is 10 then what is the value of x?

Revision 1

CHAPTER 1 to 3

1. What value does '5' have in 25,990?

a) Ten-Thousands
b) Thousands
c) Hundreds
d) Tens
e) Units

Information given below is for question 2 to 4:
Jenny drives 50 miles on Monday, 62 miles on Tuesday, 43 miles on Wednesday and 34 miles on Thursday.

2. What is the total number of miles Jenny drove on the four day period, to the nearest hundred?

a) 180
b) 189
c) 190
d) 200
e) 210

3. What is the average number of miles driven by Jenny on these days to the nearest whole number?

a) 46
b) 47
c) 48
d) 50
e) 52

4. What is the median number of miles driven?

a) 43
b) 45
c) 46.5
d) 47.5
e) 48

5. Evaluate the following: $\frac{6^2-3^2}{3^3}$

a) 27
b) 18
c) 6
d) 1
e) 0

6. Which of the following is not a factor 75?

a) 35
b) 25
c) 15
d) 3
e) 1

7. What is the remainder when 453 is divided by 3?

a) 0
b) 1
c) 2
d) 3
e) 4

8. If an even integer is added to odd integer then result is square number then which the following is true?

a) result is even
b) result is odd
c) result is zero
d) result is undefined
e) result is unknown

9. The quotient of 69,459/49 is between what numbers?

a) between 1500 to 1600
b) between 1400 to 1500
c) between 1300 to 1400
d) between 1200 to 1300
e) between 1100 to 1200

10. Evaluate 5 x 2 - 3 x 2 + 4 x 5 = ?

a) 14
b) 20
c) 24
d) 34
e) 40

11. The product of 5291 and 21 is

a) 99,191
b) 100,001
c) 101,201
d) 111,111
e) 111,351

12. Which of the following is a prime factor of 3210?

a) 6
b) 10
c) 15
d) 97
e) 107

13. Given {39, 43, 58, 61, 73, and 89} how many numbers are prime?

a) 1
b) 2
c) 3
d) 4
e) 5

14. Which of the following pairs are the next two consecutive odd numbers after 51?

a) 47 and 49
b) 48 and 49
c) 49 and 50
d) 52 and 53
e) 53 and 55

15. Which of the following number is a multiple of 5 and 9?

a) 95
b) 90
c) 59
d) 36
e) 18

16. What is the result when twenty is added to forty-five and divided by five?

a) 9
b) 10
c) 11
d) 12
e) 13

17. Evaluate the following : $\frac{(2\times 6)^2-(9+3)^2}{8^3}$?

a) 0
b) 17
c) 144
d) 288
e) 398

Information given below is for question 18 to 19:

Marcus is 24 years old, Jame is 22 years old, Barrack is 23 years old and Sam is 19 years old.

18. What is the mean age of these four people?

a) 19
b) 20.5
c) 21
d) 22
e) 23.5

19. What is the median age of these four people?

a) 19.5
b) 20.5
c) 22
d) 22.5
e) 23.5

20. Evaluate this expression (Round these number to the nearest whole number)

$$\frac{(26)^2 - (23)^2 + (16 \times 51)}{(18)^2} ?$$

a) 1.9
b) 2.5
c) 3
d) 4.7
e) 5

Chapter 4
Basic Word Problems

Steps in doing word problems:

1) Read the problem once

2) Identify the question asked

3) Choose the operation to be performed (list out the operation to be performed)

4) Solve the question carefully

Example 1

Helen makes $ 45 per day by posting ads on the internet, how much would she be making in one week period?

Question: How much does she make in one week?

Operation: $ 45 x 7 days

Answer: $ 315

Example 2

Nissan new X-trail claims that it can run 15 miles per gallon, if Jason has test driven it for 135 miles then how many gallon was used?

Question: How much fuel was used?

Operation: 135 miles ÷ 15 miles per gallon

Answer: 9 gallons

Exercise 4

1. Allen is driving his truck at a constant speed of 60 miles per hour, how many miles did he drive in 4.5 hours?

Question:

Operation:

Answer:

2. A waffle maker sells each waffle at the price of $ 3.50, if his cost is $2.25 per waffle then what is his profit per piece?

Question:

Operation:

Answer :

3. The waffle maker in the above question would have how much profit if he sells 200 waffles per day?

Question:

Operation:

Answer :

4. For an election in 2015, 3 million people registered to vote but only on a quarter actually voted. How many people went to vote?

Question:

Operation:

Answer :

5. In question 4, how many people did not go to vote?

Question:

Operation:

Answer :

6. Oranges cost $ 3 per kilogram while apple cost $ 2 per kilogram, how much would Jane have to pay if she buys 5 kilogram of each fruit?

Question:

Operation:

Answer :

7. Five people share a room in Swiss Merit hotel which cost $ 565, how much does each person pay?

Question:

Operation:

Answer :

8. Ananda makes $1,000 per month. He pays $250 for rent, $ 470 for food and save the rest. How much does he save per month?

Question:

Operation:

Answer :

9. Temperature of London noted in the morning was 13°C, in the afternoon it was 18°C and in the evening it was 11°C. What was the average temperature that day?

Question:

Operation:

Answer :

10. Mark is 42 years old, his son Pete is 16 years old. How old was Mark when his son was born?

Question:

Operation:

Answer :

Chapter 5
Advance Word Problems

Steps in doing word problems:

1) Read the problem once and underline the main idea

2) Identify the question asked

3) Choose the value and operation needed to be performed

4) Solve the question carefully

Example 1

A box contain 5 pencil cases and each case contain 20 pencils. How many pencils are in 30 boxes?

Question: Find total number of pencils in 30 boxes

Operation: 30 boxes × 5 cases × 20 pencils

             (each box has 5 cases)   (each case has 20 pencils)

Answer: <u>3000 pencils</u>

 Example 2

Jacob earns $ 50,000 per year and his boss withholds $ 2,000 for tax deduction. What is his monthly salary after deduction?

Question: How much is his salary per month?

Operation: <u>step 1</u>→ 50,000 - 2000 = $48,000 per year

                 (tax deduction)   (net salary)

        <u>step 2</u> → 48,000 ÷ 12 months =

Answer: <u>$ 4,000 per month</u>

Exercise 5

1. Mike buys the following items from the market:

> 5 kg of oranges at $ 4 per kg
>
> 3 kg of apples at $ 5 per kg
>
> 7 kg of bananas at $ 3.50 per kg

How much change would he receive if he pays with a $ 100 bill?

Question:

Operation:

Answer:

2. Paul can type 60 words in 15 mins, how long would it take him to type a report consisting of 480 words and a memo consisting of 120 words?

Question:

Operation:

Answer :

3. A group of 5 students won a raffle ticket prize of $ 590. After a tax deduction of $ 35, how much would each student receive?

Question:

Operation:

Answer :

4. For an election in 2015, 8 million people register to vote but only on a quarter actually voted. How many people did not vote?

Question:

Operation:

Answer :

5. Iphone 6 is capable of taking 35 snapshots in one minute, how many snapshots can be taken in 1 hour?

Question:

Operation:

Answer :

6. Vin Diesel claims that he can drive a quarter mile in 13.5 seconds, how many second would it take him to drive one mile?

Question:

Operation:

Answer :

7. Sean bought two jackets for $ 90 on a winter sale. The sale price of a jacket was $ 18 less than the original price quoted. What was the original price of one jacket?

Question:

Operation:

Answer :

8. Allen wants to divide a 50-acre land into building lots of 11 acres each. How many building lots will be there?

Question:

Operation:

Answer :

9. Temperature of Istanbul in the morning was 3°C, in the afternoon was 5°C hotter than morning and in the evening was 2°C colder than the morning. What is the average temperature that day?

Question:

Operation:

Answer :

10. Fred makes $ 32 per hour in regular working time and makes $ 4 more per hour in overtime rate. If he works 20 hour at regular rate and 5 hours overtime rate, then how much would he earn?

Question:

Operation:

Answer :

Chapter 6
Sequences and Scientific Notation

Sequences are **pattern** of numbers arrange in specific and predictable ways

Type of sequences	Examples
Consecutive number sequence	11, 12, 13, 14, 15 , ...
Consecutive even integer sequence	20, 22, 24, 26, 28 , ...
Multiple number sequence	5, 10, 15, 20, 25, ...
Square of integer sequence	1, 4, 9, 16, 25, 36, ...
Recurring Sequence	-1, 0, 1, -1, 0, 1, -1,...

Example 1

Find the next 3 terms of the sequence   1, 3, 7, 15, 31, __ , __, __

(Hint: Think of multiplying then adding)

Second term → (1 x 2) + 1 = 3

Third term   → (3 x 2) + 1 = 7

So, Sixth term→ (31 x 2) + 1 = <u>63</u>

Seventh term → (63 x 2) + 1 = <u>127</u>

Eighth term   → (127 x 2) + 1 = <u>255</u>

Example 2

Every month Kathy withdraw the money from saving account to pay for her tuition.
How much money will be left by the month of May?

Jan	Feb	March	April	May
$ 50,000	$ 42,000	$ 34,000		

Scientific notation is a way of shortening large numbers using power of tens

Name	Number Value	Power
Units or Ones	1	10^0
Tens	10	10^1
Hundreds	100	10^2
Thousands	1,000	10^3
Ten-Thousands	10,000	10^4
Hundred-Thousands	100,000	10^5
Million	1,000,000	10^6

Example 1

Write 5,630,000 in scientific notation.

(Hint 1: Count the number of digit and minus it by one, that is the power of ten obtained)

(Hint 2: Only one number should be in front of decimal)

5,630,000 = **5.63 x 10^6**

Example 2

Write **3.264 x 10^4** as a proper decimals

(Hint: Count the number of decimals minus it with power of tens)

3.264 x 10^4 = 32,640

Exercise 6

1. Find the next 3 terms of the sequence below

a) 3, 7, 11, 15, __, __, __

b) 15, 13.5, 12, 10.5, __, __, __

c) 45, 51, 57, __, __, __

a) 1, 1, 2, 3, 5, 8, __, __

b) 0, 1, 4, 9, __, __

c) 2, 6, 12, 20, __, __

d) 1000, 500, 250, __, __

2. Find the 8th term of this sequence: 2, 5, 10, 17, _____

3. Find the formula for each sequences:

a) 5, 10, 15, 20

b) 14, 17, 20, 23

c) 99, 90, 81, 72

4. Every month Max deposit the money into saving account. How much money will be there by the month of May?

Jan	Feb	March	April	May
$ 124	$ 184	$ 244		

5. Express each number in scientific notation

a) 620,000   =

b) 32,000   =

c) 1,990   =

d) 9,000,000   =

e) 888,000,000 =

6. Express each into standard form

a) 2×10^3 =

b) 312×10^4 =

c) 9555×10^3 =

d) 881×10^7 =

e) 20123×10^6=

Revision 2

CHAPTER 4 to 6

1. Xiao-mi is paid $35 per day for her work at iStudio and a commission of $10 for selling one iPhone. What would her pay be if she work for 15 days and sold 23 iPhones?

a) 855
b) 755
c) 675
d) 525
e) 230

Information given below is for question 2 to 4:
The distance from Earth to the Sun is called an astronomical unit, or AU, which is used to measure distances throughout the solar system. One AU has been defined as 150,000,000,000 meters.

2. Which of the following expresses one Au value in scientific notation?

a) 150×10^9 m
b) 15×10^9 m
c) 15×10^{10} m
d) 15×10^{11} m
e) 15×10^{11} m

3. Which of the following expresses half of the distance between the Earth and the Sun in scientific notation?

a) 75×10^9 m
b) 75×10^9 m
c) 75×10^{10} m
d) 75×10^{11} m
e) 075×10^{11} m

4. Jupiter is about 5.2 AU from the Sun, which of the following represents the distance in meters between them?

a) 752×10^9 m
b) 752×10^9 m
c) 78×10^{11} m
d) 72×10^{11} m
e) 754×10^{12} m

5. Big Momma Jewelry store pays 3 salesman at the rate of $12 per hour and 5 goldsmiths at the rate of $25 per hour. If they work 30 hours per week, how much does the store pay them?

a) 161
b) 1610
c) 4830
d) 6440
e) 8880

6. Thai-Elephant foundation needed to raise 350,000 baht to help save the handicap elephants by selling amulets. Jeff sold 99 amulets, Mark sold 53 amulets and Anya sold 48 amulets.   If everyone sold the amulets at the same price, how much would one amulet cost?

a) 175
b) 200
c) 550
d) 1500
e) 1750

7. What would be the next number in the sequence 6, 13, 27, 55, ?

a) 85
b) 90
c) 101
d) 111
e) 112

8. What is the formula to find the nth term of the sequence 12, 17, 22, 27,... ?

a) 5n + 12
b) 5n + 7
c) 5 + 12(n-1)
d) 12n + 1
e) 12n + 5

9. This year Paul will be 3 years older than Frank and Frank will be 5 years younger than Charlie. If Charlie is 9 years old then how old would Paul be this year?

a) 7
b) 9
c) 11
d) 15
e) 17

10. Fibonacci sequence {1,1,2,3,5,...} is generated by adding the two consecutive term to obtain the next term, for instance if we want the fourth term we add the third and second term. What is the eighth term of this sequence?

a) 8
b) 13
c) 21
d) 34
e) 55

11. Which of the following pairs are the next two consecutive numbers of the sequence $11a, 9a^2, 7a^3$? (arranged in order)

a) $5a^3$ and $3a^4$
b) $5a^4$ and $3a^3$
c) $5a^4$ and $3a^5$
d) $3a^4$ and $5a^5$
e) $3a^3$ and $5a^4$

12. Which of the following is the tenth term of the sequence that has the nth term of $7n + 2$?

a) 17
b) 19
c) 70
d) 72
e) 87

13. A car travel 800 km at an average speed of 60 km/hr., how many hours and minutes does it take to travel this distance?

a) 13 hours and 20 minutes
b) 13 hours and 33 minutes
c) 13 hours and 44 minutes
d) 14 hours and 20 minutes
e) 14 hours and 33 minutes

14. At Trailly Textile Company 63 employees walk to work, 42 drive to work and 91 use public transportation. If there are 500 employees, how many travel by other means?

a) 169
b) 196
c) 246
d) 304
e) 395

Information given below is for question 15 to 17:

Store	Table	Chairs
Roony's	$ 56	$ 12 each
Beckies'	$ 65	$ 20 for 2

15. Find the cost of a table and 12 chairs at Roony's?

a) 144
b) 156
c) 185
d) 200
e) 305

16. Find the cost of a table and 12 chairs at Beckies'?

a) 144
b) 156
c) 185
d) 200
e) 305

17. If Roony's is offering a discount then how much money would you save compared to Beckies' if you buy a table and 12 chairs?

a) 17
b) 18
c) 65
d) 105
e) not enough information is given

18. The distance from the Earth to the nearest NASA probe is about 752×10^{9} km, which of the following represents this in a whole number?

a) 75,200,000
b) 752,000,000
c) 7,520,000,000
d) 75,200,000,000
e) 752,000,000,000

19. Jamie Foxx wanted to reduce his weight by a least 3 pounds per week for his new movie. If he weighs 210 pounds, how many weeks would it take him to reduce his weight to 180 pounds?

a) 5
b) 10
c) 20
d) 30
e) not enough information is given

20. Putin drives to work every morning and park his car at Siam Paragon, the gasoline cost him 300 baht per week and the parking cost him 60 baht per day. If he works 5 days per week, how much does it cost him daily to go to work?

a) 60
b) 72
c) 100
d) 120
e) not enough information is given

Chapter 7
Decimals Multiplication and Division

Rules in multiplying decimals:

1. Multiply the two numbers by ignoring decimals point

2. Count the decimals point of the first number and the second number then add them

3. Move the decimals point forward from the result in step one by the addition result of step two

Example 1

Evaluate   23.25 x 0.002 = ?

Step 1:   Step 2:

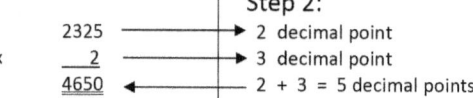

Step 3:

Move the decimal points for 4650 five times

23.25 x 0.002 = 0.04650 = <u>0.0465</u>

Example 2

Evaluate   23.25 x 1000 = ?

Step 1:   Step 2:

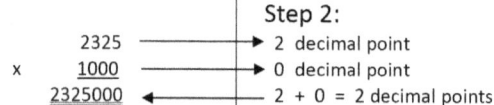

Step 3:

Move the decimal points for 2325000 two times

23.25 x 1000 = 23250.00 = <u>23250</u>

Rules in dividing decimals:

1. The divisor should contain no decimal point, if there are decimal points then get rid of it !

2. Get rid of divisor decimal points by multiplying it with $10^{decimal\ points}$

3. Just divide normally

Example 3

Evaluate   23.25 ÷ 5 = ?

Since there are no decimal point at the divisor, so we can just divide them normally

$$
\begin{array}{r}
4.65 \\
5\overline{)23.25} \\
\underline{20}\downarrow \\
32 \\
\underline{30}\downarrow \\
25 \\
\underline{25} \\
0
\end{array}
$$

23.25 ÷ 5 = <u>4.65</u>

Example 4

Evaluate   23.25 ÷ 0. 5 = ?

Since there is one decimal point at the divisor, so we multiply both the numbers by 10^1

(23.25 x 10) ÷ (0. 5 x 10) = 232.5 ÷ 5

$$
\begin{array}{r}
46.5 \\
5\overline{)232.5} \\
\underline{20}\downarrow \\
32 \\
\underline{30}\downarrow \\
25 \\
\underline{25} \\
0
\end{array}
$$

23.25 ÷ 0.5 = <u>46.5</u>

Exercise 7

1. Evaluate the following product

    a) 5.3×0.9 =

    b) 0.234×0.12 =

    c) 7.05×0.002 =

    d) 80.2×0.25 =

2. Evaluate the following product

    a) 1.902×10 =

    b) 0.03×100 =

    c) 1000×5.06 =

    d) 0.0219×10000 =

3. Evaluate the following quotient

    a) $3.42 \div 2$ =

    b) $69.3 \div 3$ =

    c) $15.5 \div 5$ =

4. Evaluate the following quotient

    a) $5.4 \div 100$ =

    b) $32.55 \div 1000$ =

    c) $2.52 \div 10$ =

5. Evaluate the following quotient

    a) $8.6 \div 0.2$ =

    b) $2.7 \div 0.03$ =

    c) $12.34 \div 0.005$ =

Chapter 8
Fractions

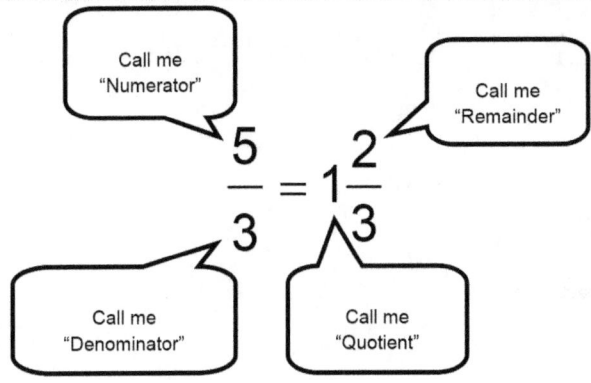

Rules in adding/subtracting fractions:

1. Find the 'common denominator'

2. Make the fractions have equal denominator

3. Add/subtract the fraction (only the numerator part)

Example 1

Evaluate $\dfrac{3}{5} + \dfrac{4}{9}$ = ?

Step 1:

Common Denominator = 5 x 9
= 45

Step 2:

$$\frac{3}{\boxed{5}} + \frac{4}{\textcircled{9}} = \frac{3 \times \textcircled{9}}{5 \times \textcircled{9}} + \frac{4 \times \boxed{5}}{9 \times \boxed{5}}$$

Step 3:

Just add the numerator now

$$\frac{27}{45} + \frac{20}{45} = \frac{27 + 20}{45} = \frac{47}{45} = 1\frac{2}{45}$$

Example 2

Evaluate $\qquad 4\frac{3}{5} - 2\frac{1}{10} = ?$

Step 1:

Common Denominator = 10
(5 and 10 gives a common of 10)

Step 2:

$$\frac{[(4\times 5)+3]\times 2}{5\times 2} - \frac{(2\times 10)+1}{10} =$$

Step 3:

Just subtract the numerator and simplify to lowest term

$$\frac{46}{10} - \frac{21}{10} = \frac{46-21}{10} = \frac{25}{10} = \frac{5}{2} = 2\frac{1}{2}$$

Rules in multiplying fractions:

Make sure to change mixed numbers to improper fraction

Example 3

Evaluate $\qquad 4\frac{3}{5} \times 2\frac{1}{2} = ?$

Step 1:
Change 'mix number' to 'proper fraction'

$$\frac{[(4 \times 5) + 3]}{5} \times \frac{(2 \times 2) + 1}{2}$$

Step 2:
Just multiply directly

$$= \frac{23}{5} \times \frac{5}{2} = \frac{23 \times 5}{5 \times 2} = \frac{23}{2} = 11\frac{1}{2}$$

Rules in dividing fractions:

1. Change the division(÷) sign to multiplication(×) then swap the numerator with the denominator

of the divisor term

2. Multiply the numerator directly with each other and also the denominator

Example 4

Evaluate $\qquad \frac{13}{7} \div \frac{39}{14} = ?$

Step 1:
Change '÷' to '×' then *swap*

$$\frac{13}{7} \div \frac{39}{14} = \frac{13}{7} \times \frac{14}{39}$$

Step 2:
Just multiply directly

$$= \frac{13 \times 14}{7 \times 39} = \frac{2}{3}$$

Exercise 8

1. Evaluate the following sum

    a) $\frac{2}{5} + \frac{3}{5}$ =

    b) $3\frac{1}{2} + 9\frac{2}{3}$ =

    c) $\frac{3}{4} + \frac{2}{5} + \frac{7}{10}$ =

    d) $10\frac{1}{4} + 3\frac{2}{3} + 2\frac{5}{6}$ =

2. Evaluate the following differences

    a) $\frac{2}{3} - \frac{3}{5}$ =

    b) $5\frac{3}{7} - 2\frac{2}{7}$ =

    c) $\frac{3}{4} - \frac{1}{10} - \frac{3}{20}$ =

    d) $6\frac{1}{3} - 1\frac{2}{3} - 3\frac{2}{9}$ =

3. Evaluate the following product

    a) $\frac{2}{3} \times \frac{5}{11}$ =

    b) $4\frac{1}{2} \times 12\frac{1}{3}$ =

    c) $\frac{3}{4} \times \frac{2}{21} \times \frac{7}{10}$ =

    d) $5\frac{5}{7} \times 2\frac{1}{3} \times 8\frac{1}{6}$ =

4. Evaluate the following quotient

    a) $\frac{12}{11} \div \frac{6}{55}$ =

    b) $4\frac{1}{2} \div 12\frac{1}{3}$ =

    c) $\frac{11}{9} \div \frac{22}{45} \div \frac{3}{27}$ =

    d) $4\frac{3}{5} \div 2\frac{1}{23} \div 1\frac{5}{6}$ =

Chapter 9
Converting Fraction↔Decimal

Example 1

Convert $\frac{3}{5}$ to decimals

$$
\begin{array}{r}
0.6 \\
5\overline{)3.0} \\
\underline{0}\downarrow \\
30 \\
\underline{30} \\
0
\end{array}
$$

Therefore $\frac{3}{5} = \underline{0.6}$

Short Cut (change denominator to base 10)

Change $\frac{3}{5} \rightarrow \frac{3}{5} \times \frac{20}{20} = \frac{60}{100} = \underline{0.60}$

Example 2

Convert $\frac{13}{3}$ to decimals

$$
\begin{array}{r}
4.333 \\
3\overline{)13.000} \\
\underline{12}\downarrow \\
10 \\
\underline{9}\downarrow \\
10 \\
\underline{9}\downarrow \\
10 \\
\underline{9} \\
1
\end{array}
$$

now we know it's a repeating decimals

Therefore $\frac{13}{3} = \underline{4\overline{3}}$

To convert decimals to fraction:

Step 1: Remove decimals by multiplying both numerator and denominator with 10 to the power of decimal place

Step 2: Reduce the fraction to lowest form

Example 3

Convert   0.25 to fraction

Step 1:

Remove the decimals by multiplying numerator and denominator of 0.25 (2 d.p.) with 10^2 or 100

$$\frac{0.25}{1} \times \frac{100}{100} = \frac{25}{100}$$

Step 2:

Reduce this fraction to lowest form

$$\frac{25}{100} = \frac{5}{20} = \frac{1}{4}$$

Example 4

Convert   3.125 to fraction

Step 1:

Remove the decimals by multiplying numerator and denominator of 3.125 (3 d.p.) with 10^3 or 1000

$$\frac{3125}{1} \times \frac{1000}{1000} = \frac{3125}{1000}$$

Step 2:

Reduce this fraction to lowest form

$$\frac{3125}{1000} = \frac{25}{8} = 3\frac{1}{8}$$

Scientific notation with small numbers help shortening long decimals

Name	Number Value	Power
Units or Ones	1	10^0
Tenths	0.1	10^{-1}
Hundredths	0.01	10^{-2}
Thousandths	0.001	10^{-3}
Ten-Thousandths	0.0001	10^{-4}
Hundred-Thousandths	0.00001	10^{-5}
Millionths	0.000001	10^{-6}

Example 1

Write 0.00632 in scientific notation.

(Hint 1: Count the number of digit after decimal until the first significant number, that is the power of ten with negative obtained)

(Hint 2: Only one number except zero should be in front of decimal)

$0.00632 = \underline{6.32 \times 10^{-3}}$

Example 2

Write 2.9031×10^{-4} as a proper decimals

(Hint: Look at the negative power, add one to it → you will get -3 but take it as 3 and write 3zeros in front of 2)

$2.9031 \times 10^{-4} = \underline{0.00029031}$

Exercise 9

Question	Fraction	Decimals
1	$\dfrac{12}{5}$	
2		0.6
3	$\dfrac{3}{10}$	
4	$\dfrac{19}{10}$	
5		99.999
6	$\dfrac{5}{3}$	
7	$8\dfrac{1}{10}$	
8	$25\dfrac{1}{3}$	
9		8.25
10		0.121212

11. Express each as scientific notation

    a) 0.0005345

    b) 0.000000940

    c) 0.00001010

    d) 0.02

12. Write each scientific notation as decimals

    a) 314×10^{-2}

    b) 226×10^{-6}

    c) 604×10^{-3}

    d) 9014×10^{-5}

1. Paula drove 320 miles on 9 gallons of gasoline on her Toyota Prius. To the nearest tenth, what is her average fuel consumption in miles per gallon?

a) 35.5
b) 35.56
c) 35.6
d) 36
e) 40

Information given below is for question 2 to 4:
The total amount for the budget for this month was $ 1,400.

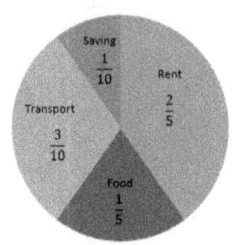

Monthly Expenses Budget

2. How much was spent on rent?

a) 140
b) 280
c) 360
d) 420
e) 560

3. How much was spent on food and transport?

a) 420
b) 540
c) 560
d) 600
e) 700

4. How much more money was spent on food and transport than rent?

a) 140
b) 280
c) 360
d) 420
e) 560

5. Evaluate 25.5 ÷ 0.15?

a) 0.17
b) 1.7
c) 17
d) 170
e) 1700

6. Evaluate 6.1^2

a) 36.31
b) 37.21
c) 38.81
d) 38.91
e) 39.81

7. Evaluate $0.1^2 \div 0.01^3$?

a) 10
b) 100
c) 1000
d) 10000
e) 100000

8. Size of a type C bacteria is about 0.00000000324 m long, express this in scientific notation.

a) 3.24×10^{-8}
b) 3.24×10^{-9}
c) 3.24×10^{-10}
d) 32.4×10^{-8}
e) 32.4×10^{-9}

9. Which of the following lists are numbers arranged from least to greatest?

a) $0.55, \frac{3}{5}, 0.005, 0.05, \frac{1}{2}$
b) $0.55, \frac{3}{5}, \frac{1}{2}, 0.005, 0.05$
c) $0.05, \frac{3}{5}, \frac{1}{2}, 0.005, 0.55$
d) $0.005, 0.05, \frac{3}{5}, 0.55, \frac{1}{2}$
e) $0.005, 0.05, \frac{1}{2}, 0.55, \frac{3}{5}$

10. Write 9.989×10^{-3} as a normal number

a) 9989
b) 99.89
c) 0.09989
d) 0.009989
e) 0.0009989

11. Which of the following best represents the decimals value of $\frac{7}{9}$?

a) 0.8
b) $0\overline{7}$
c) 0.78
d) 0.778
e) 0.789

12. Evaluate $1\frac{2}{5} + 3\frac{2}{3} - 2$

a) $3\frac{3}{5}$
b) $3\frac{7}{15}$
c) $3\frac{1}{15}$
d) $4\frac{1}{15}$
e) $4\frac{7}{15}$

13. Evaluate $\frac{2}{5} \div 2\frac{4}{15} \times 17$

a) 0.3
b) 0.27
c) 0.17
d) 0.05
e) 0.015

14. Maria bought 3.8 pound of cheese at $10.20 per pound, if she pays a 50 dollar bill how much change should she receive?

a) 10.16
b) 11.24
c) 12.80
d) 24.50
e) 38.76

15. One pound equals 0.45 kilogram, Frank weighs 230 pounds. What is Frank's weight in kilograms ?(rounded to the nearest units)

a) 103
b) 104
c) 105
d) 108
e) 110

16. Evaluate $3.2 \times 10^{-2} \times 5 \times 10^3$?

a) 0.16
b) 1.6
c) 16
d) 160
e) 1600

17. Johnie Bravo wanted to reduce his weight by 3 pounds per week for his school prom. If he weighs 228 pounds, how many weeks would it take him to reduce his weight to 180 pounds?

a) 16
b) 18
c) 24
d) 30
e) not enough information is given

18. One-fourth of student at AIMS iGen drive their car to study, two-third take BTS and the rest walk. If there are 240 students, how many walk to AIMS?

a) 20
b) 40
c) 60
d) 160
e) not enough information is given

Information given below is for question 19 to 20:

Type of Cheese	Price per oz.	Price per 10 oz.
Mozzarella	$ 0.35	$ 2.89
Cheddar	$ 0.66	$ 5.25

19. Mark wants to buy 12 oz. of Mozzarella and 27 oz. of Cheddar cheese, what is the cheapest price he must pay?

a) 14.43
b) 18.71
c) 22.02
d) 32.54
e) 34.22

20. If Mark pays $50 bill, how much change would he expect to get?

a) 35.57
b) 31.29
c) 27.98
d) 17.46
e) 15.78

Chapter 10
Ratio and Proportion

Ratio is a way of comparing numbers by division.
Ratio can be represented by colon (:), the word to, or as a fraction.
Always reduce ratio to its lowest term

For example

If there are 30 men and 20 women in a class. We could represent the ratio of men to women in two ways.

Ratio form → men : women = 30:20	Fraction form → men : women = $\dfrac{30}{20}$
Reduced Ratio form → men : women = 3:2	Reduced Fraction form → men : women = $\dfrac{3}{2}$

So we could say the ratio of men to women is,

men: women $= 3$ to $2 = 3:2 = \dfrac{3}{2}$

Example 1

Billy makes $3,000 per month, she pays $ 500 for rent.

a) What is the ratio of her rent to her income?

Rent : Income

500 : 3,000

$=$ 1 : 6

b) What is the ratio of her income to her rent?

Income : Rent

3,000 : 500

$=$ 6 : 1

Example 2

Find the value of x, if x : 5 = 12 : 15

$$\frac{x}{5} = \frac{12}{15}$$

$$x = \frac{12}{15} \bullet 5$$

$$x = 4$$

Example 3

In NBA game stats Kobe Byrant misses 3 shots for every 8 shots that he made in basket. All together he shoots 55 shots, how many shot did he made in basket?

Total shot made = 3 miss + 8 made = 11 shots

Now take ratio of

shot made : total shot → 8 : 11 = M : 55

$$\frac{8}{11} = \frac{M}{55}$$

$$M = \frac{8 \times 55}{11}$$

$$M = 40 \quad \text{shots made}$$

Exercise 10

1. Simplify each ratio

    a) 24 : 32

    b) 8.1 : 4.5

    c) $\dfrac{102}{105}$

    d) \$ 144 : \$ 720

2. Find the missing term

    a) 8 : w = 4 : 15

    b) 1.5 : 4.5 = 10.5 : y

    c) $\dfrac{2}{11} = \dfrac{n}{77}$

    d) $\dfrac{3}{18} = \dfrac{2}{x}$

    e) $\dfrac{4x}{9} = \dfrac{10}{15}$

3. Joe made \$ 40 per day, save \$ 5 in his piggy bank and spent the rest.

    a) What is the ratio of the amount he saved to amount he made?

    b) What is the ratio of the amount he made to amount he saved?

    c) What is the ratio of the amount he spent to amount he made?

4. A GED class consists of 72 students and 12 of the students are female.

    a) What is the ratio of male students to female students?

    b) What is the ratio of female students to male students?

    c) What is the ratio of male students to total students?

    d) What is the ratio of total students to male students?

5. Paula earns \$40 a day and she takes home \$25. Paula's gross pay in this month is \$800, how much does she takes home?

7. Intel produces a cpu at the assembly line with the ratio of good to defective parts as 30:1. Each day the company produces 15,500 parts. How many of these parts will be defective?

8. Mango cost \$ 3 per dozen, how much do 5 mangoes cost?

Chapter 11
Percentage

Percent means out of hundred, this can be converted to fraction or decimals.

$100\% = \dfrac{100}{100} = 1$

one whole is 100%

$a\% = \dfrac{a}{100}$

Example 1

If there are 30 men and 20 women in a class

    a) What percent are women?

$$\text{\% of women} = \frac{\text{women}}{\text{total}} \times 100\% = \frac{20}{50} \times 100\% = \underline{40\%}$$

    b) What percent are men?

$$\text{\% of men} = \frac{\text{men}}{\text{total}} \times 100\% = \frac{30}{50} \times 100\% = \underline{60\%}$$

Conversion of Percentage to decimals or fractions

To convert percent to decimals just move the decimal point 2 times to the left	*To convert percent to fraction just divide the percent by 100 (also reduce it to lowest form)*
23.0 % = 0.23 60.5 % = 0.605	$23.0\% = \dfrac{23}{100}$ $16\% = \dfrac{16}{100} = \dfrac{4}{25}$

Example 2

A dress originally priced at $ 150 is on sale for 30% off. If Marsha wanted to buy this dress, then how much does she have to pay?

30% off → means she has to pay 70% of the price (100% - 30% = 70%)

Payment = 70% of 150

$$= \frac{70}{100} \times 150 \; = \; \$\,105$$

So Marsha is paying $ 105, she saved $ 45.

Example 3

Lilo bought a new Samsung Galaxy 8 for $ 110 on sale, which was originally priced at $ 200. What was the percent discount he received from the shop?

Save = 200 - 110 = $ 90

$$\text{Percent discount} \; = \; \frac{\text{Save}}{\text{Original Price}} \times 100\%$$

$$= \frac{90}{200} \times 100\%$$

$$= 45\,\%$$

Lilo got 45 % discount on his Samsung Galaxy 8

Exercise 11

1. Convert the following to fractions and decimals

Percentage	Decimals	Fractions
30%		
20%		
10%		
5%		
1%		
200%		
25%		
75%		

2. Evaluate the following:

      a) What is 5 % of 700

      b) Find 12 % of 95

      c) 7 is what percent of 49?

      d) Find 2.5 % of 120

      e) What is 400 % of 6

3. Ann earns a salary of $ 5,200 per month. She spends 33 % on her rent, how much is her rent?

4. In a GED class 96 students took the exam, 72 people passed the exam. What percent of the students did not pass?

5. Tony bought a second hand Ferrari for $ 20,000 which included a sale tax of 6%. What is the price of this car without a sale tax?

6. BigC gives a special offer for new year, if you buy worth 2,000 baht you get 5 % discount and if you buy more than 2,000 baht worth you get an extra 2% off from the discounted price. If Nick plans to shop for 3,500 then how much money could he save?

Chapter 12
Percent Change and Interest Rate

Percent Change means *Percent Increase* or *Percent Decrease*

$$\%Change = \frac{(New-Original)\times 100\%}{Original}$$

Example 1

Maria weighs 96 lbs. on Monday and after partying too much for the entire week
she discovered that on Saturday her weight was 101 lbs. What is the percent increase in her
weight?

$$\%Change = \frac{(New-Original)\times 100\%}{Original}$$

$$= \frac{(101-96)\times 100\%}{96}$$

$$= 5.2\ \%$$

The percent increase in her weight is   5.2 %

Example 2

A dress was bought for $ 48 in a sale. What is the original price of this dress if it was sold at a
discount of 20%?

20% off → means we pay 80% of the price *(100% - 20% = 80%)*

$ 48 = 80% of Original

$$\$\ 48 = \frac{80}{100}\times Original$$

$$Original = \frac{100}{80}\times 48$$

$$Original = \$\ 60$$

The original price of this dress was $ 60

Example 3

Jordan sells comic book at Red-Cross fair. He will markup the price of each book by 18% before selling it. If the cost of each book is $ 5, then what would be the retail price?

Retail price = Cost + Markup(or Profit)

Retail price = $ 5 + 18% of $ 5

Retail price = $ 5 + (0.18 x $ 5)

Retail price = $ 5.90

Jordan should sell his book for $ 5.90 each

Simple interest is the amount of money one receives after depositing money for certain period of time at a certain rate.

$$Interest = \frac{(Princi pa Rat ex Ti me)}{100}$$

Example 4

Karl deposited $ 12,000 in his saving account which pays 2.5 % simple interest annually. If he deposited the money for 6 years, how much money will he have at the end of this time period?

Principal = $ 12,000

Rate = 2.5 %

Time = 6 years

$$Interest = \frac{(Princi pa Rat ex Ti me)}{100} = \frac{(12000 \times 25 \times 6)}{100}$$

Interest = $ 1,800

Total money = Principal + Interest = 12,000 + 1,800 = $ 13,800

In the end Carl will have $ 13,800 in his account.

Exercise 12

1. Sam has paid his mother $ 4,500, which is 35% of the total amount he borrowed from his mother to purchase an apartment. How much money did Sam borrow from his mother?

2. AIS charges 2 baht per minute for voice call, while DTAC charges 1.5 baht per minute. What is the percent difference between the two carriers?

3. Shin Corp. bought a building in 2000 for 35million baht. They sold it in 2014 for 140million baht. Find the rate of increase in the value of the building.

4. Jolly's gross salary is 25,000 baht a month. His boss withholds 5% for government tax, 8% for social security, and 7% for provident fund. What is Jolly's net salary?

5. A goldsmith adds 12.5 % profit to the selling price of gold in Chinatown. If today the gold is sold at 22,500 baht then what would be the cost of the gold for this goldsmith?

6. Of the 150 people who were scheduled to go for an excursion with AIMS, 30% did not show up. Of these 'no-shows' one-third were from AIMS Phayathai branch. How many people from AIMS Phayathai did not show up?

7. Mr. Arthur bought a Rolex watch from Switzerland for $ 4,000 and sold it to his client in Bangkok for $ 5,500. What was his percent profit in this deal?

8. Thirty-five percent of all the road accidents are caused by mobile phone. Of these, 20% involve chatting while driving. Out of 50,000 car accident per month, how many are caused by using mobile phone for chatting?

9. Elly drives her car at the rate of 60 mph, if she increases her speed by fifty percent she would reach her destination in 1.5 hours. How many miles long is her destination?

Revision 4

CHAPTER 10 to 12

1. The blueprint is drawn with a scale of 1.5 inches to 9 ft. on the plan. A house is 21 inches long, what is the actual distance?

a) 14
b) 31.5
c) 35
d) 89.5
e) 126

Information given below is for question 2 to 4:

Fred work as a programmer at JAVA company, his starting salary is $ 36,000 per year. Every year his company will raise his salary by 10%. The company also gives a bonus of $ 2,500 at the end of the year, given that his performance is good.

2. How much does Fred made altogether at the end two year, if his performance was poor?

a) 39600
b) 75600
c) 79200
d) 80600
e) 84200

3 How much does Fred made at the end two year, if his performance was good?

a) 39600
b) 75600
c) 79200
d) 80600
e) 84200

4. How much more would Fred earn in the third year than in the first year of work, given that his performance was poor?

a) 3600
b) 7560
c) 11916
d) 12578.5
e) 47916

5. Evaluate 0.345:0.0005?

a) 690
b) 69
c) 6.9
d) 0.69
e) 0.069

6. What is 30% of 60% of 900?

a) 36
b) 72
c) 128
d) 162
e) 270

7. Mark bought a shirt for $ 40 which was on sale of 60%. What was the original cost of this shirt?

a) 50
b) 66.7
c) 80
d) 100
e) 125

12. Sam bought a Porsche in 2010 for   5

8. Alice fills a cleaning solution with 11 cups of water and add 1 cup of bleach. If she needs to use 3 gallons of cleaning solution, how many cups of bleach should she use? (1 gallon = 16 cups)

a) 3
b) 4
c) 6
d) 8
e) 12

9. Karman can shoot a free-throw on 65% of 120 shots, how many ball made a basket?

a) 78
b) 87
c) 98
d) 102
e) 113

10. Find the value of x if x : 5 = 54 : 90

a) 3
b) 6
c) 12
d) 18
e) 21

11. Eminem bought a pack of M&M. There are color coated chocolates which contain 5 red, 4 blue, 7 green and 4 brown. What fraction of chocolates are red?

a) 1/2
b) 1/3
c) 1/4
d) 1/5
e) 1/7

million baht, if he sold it today it would worth 3 million baht. By what percent did the price go down?

a) 30%
b) 35%
c) 40%
d) 45%
e) 48%

13. A bank offer an annual interest of 5% on deposits. If you deposit $ 8,900 for 10 years how much money will you have?

a) 445
b) 4450
c) 9345
d) 13350
e) 15420

14. Ryan bought 2.5 pound of cheese at $10.5 per pound in July, two months later he bought 2.5 pound of cheese at $11.55 per pound. By what percent did the price increase?

a) 3%
b) 5%
c) 10%
d) 12%
e) 25%

15. Shrek makes $2429 a month and pays $ 605 a month for his rent. Rent is approximately what percent of his income?

a) 3%
b) 5%
c) 10%
d) 12%
e) 25%

16. Which of the following is equal to 1/2?

a) 90%
b) 65%
c) 55%
d) 50%
e) 30%

17. For every 3 days that it rains 2 days it will snow. In a 30 day month, how many days will it snow?

a) 10
b) 12
c) 18
d) 20
e) not enough information is given

18. One-fourth of student at AIMS iGen drive their car to study, one-fifth took BTS and the rest walk. What percent walk to AIMS ?

a) 20%
b) 40%
c) 45%
d) 50%
e) 55%

Below information is for question 19 to 20:

Shop	Cost of bicycle	Discount
Adam	$ 1,239	10 %
Bally	$ 1,309	35 % off for second bicycle

19. Joe and his friend wanted to buy bicycles and they decided to buy from Bally. How much would each bicycle cost if they pay equal amount?

a) 850.85
b) 900.85
c) 1079.93
d) 1178.1
e) 1235.25

20. How much money has Joe saved (approximately) by buying from Bally's than Adam's?

a) 25
b) 35
c) 45
d) 65
e) 85

Chapter 13
Interpreting Scales and Graphs

Scale is a way of representing a value of measurement.
Graph can be used to represent any kind of information in the form of bar graph, pie chart or line graph.

Example 1

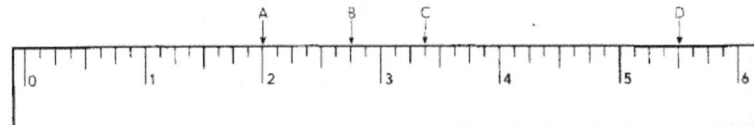

If all the length in the ruler are in inches, find the following :

a) What is the distance from A to C?

$$A = 2 \text{ in}$$

$$C = 3\frac{3}{8} \text{ in}$$

Distance from A to C = C - A = $3\frac{3}{8}$ - 2 = $1\frac{3}{8}$ in

b) What is the distance from B to D?

$$B = 2\frac{6}{8} \text{ in}$$

$$D = 5\frac{4}{8} \text{ in}$$

Distance from B to D = D - B = $5\frac{4}{8}$ - $2\frac{6}{8}$ = $2\frac{6}{8}$ in

Example 2

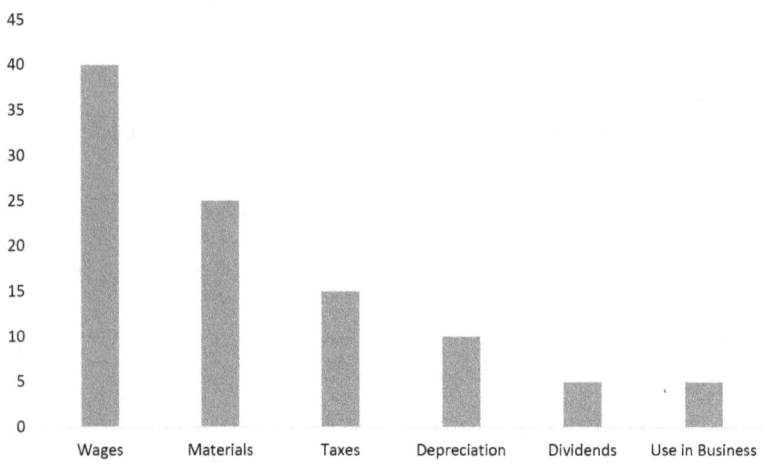

The graph above shows the expenses at PTT company (scales are in percentage)

a) What percent is spent on taxes?

        Based on the graph 15% is spend on taxes

b) If the total spending was 30million baht then how much was spent on materials?

        Based on the graph 25% are spent on materials

        Spending on material   = 25% of 30million

$$= \frac{25 \times 30{,}000{,}000}{100}$$

        = 7,500,000 baht

Total spending on material is 7,500,000 baht

Exercise 13

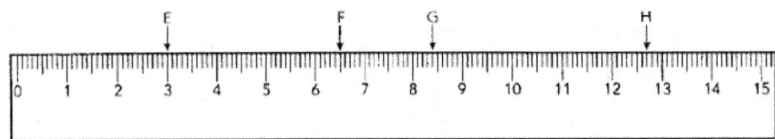

Use the figure above for question 1 to 4

1. What is the distance from E to F?

2. What is the distance from G to H?

3. How much longer is EF than FG?

4. GH is what fraction of EH?

5. Information below represents the expense in Prayuth family. This family makes an average of 200,000 baht per month

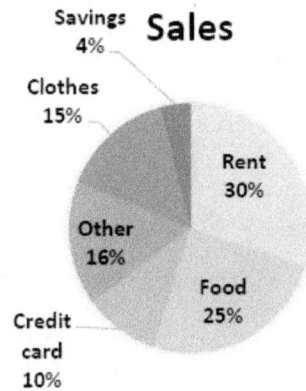

Sales

a) How much was spent on rent?

b) How much more was spent on food than clothes?

c) How much money would they save in one year?

d) How much was spent on food and credit card?

6. The information below represents the paid attendance record of Panthers and Bears baseball teams. The graph represents attendance of fans over 10 year period

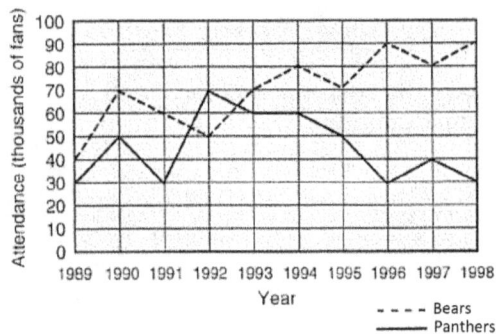

a) In what year did the Panthers' attendance exceed the Bears?

b) In 1997 how many more people attended the Bears games than the Panthers games?

c) By what percent did the Panthers attendance decrease from 1992 to 1996?

d) By what percent did the Bears attendance increase from 1992 to 1996?

7. The graph below represents the birthday of a group of GED students in IGENs.

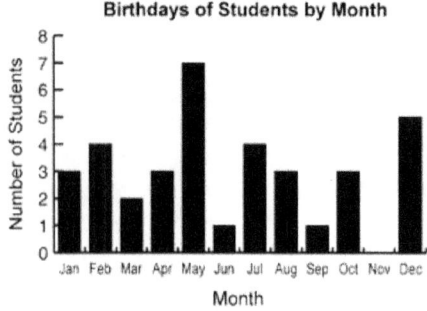

a) How many students were born in August?

b) What is the total number of student?

c) What fraction of students were born in December?

d) How many more student were born in May than in February?

Chapter 14
Converting Units

Unit is a mean of measurement that gives value to mass, length or volume
To convert unit we will be using the concept of ratio

Customary Units of Measure

Measures of Length	Measures of Weight
1 foot (ft.) = 12 inches (in.) 1 yard (yd) = 36 inches 1 yard = 3 feet 1 mile (mi) = 5280 feet 1 mile = 1760 yards	1 pound (lb.) = 16 ounces (oz.) 1 ton (T) = 2000 pounds
Liquid Measures	Measures of Time
1 pint (pt) = 16 ounces (oz.) 1 cup = 8 ounces 1 pint = 2 cups 1 quart (qt) = 2 pints 1 gallon (gal) = 4 quarts	1 minute (min) = 60 seconds(sec) 1 hour (hr.) = 60 minutes 1 day = 24 hours 1 week (wk) = 7 days 1 year (yr) = 365 days

Metric Units of Measure

Measures of Length	Measures of Weight
1 meter (m.) = 1000 millimeters(mm.) 1 meter = 100 centimeter (cm.) 1 meter = 10 decimeter (dm.) 1 kilometer (km.) = 1000 meter	1 gram (g.) = 1000 milligrams (mg.) 1 kilogram (kg.) = 1000 gram
	Measures of Liquid
	1 liter (L) = 1000 milliliter (ml.)

Interchanging Measures

Lengths	Weights
1 inch = 2.54 centimeters 1 foot = 0.3048 meter 1 mile = 1.6 kilometers	1 pound = 0.453 kilogram 1 kilogram = 2.2 pounds

Example 1

12 yards = _____ feet

we know → 1 yard = 3 feet

take ratio → $\dfrac{1\,yd}{3\,ft} = \dfrac{12\,yd}{?\,ft}$

solve → ? ft. $= \dfrac{12 \times 3}{1}$ = 36 ft.

So 12 yd is equal to 36 ft.

Example 2

Convert 3560 meter to kilometer

we know → 1 kilometer = 1000 meter

take ratio → $\dfrac{1\,km}{1000\,m} = \dfrac{?\,km}{3560\,m}$

? km $= \dfrac{3560}{1000}$ = 3.560 km

So 3560 m is equal to 3.56 km

Example 3

How many ounces equal one gallon?

we know →

1 gallon = 4 quarts
1 quart = 2 pints
1 pint = 16 ounces

solve → 1 gallon = 1 x 4 quarts x 2 pints x 16 ounces

→ 1 gallon = 128 ounces

Therefore one gallon is equal to 128 oz.

Exercise 14

1. 3 day = _____ hours

2. 2.5 pound = _____ ounces

3. 8 inches = _____ foot

4. 3 yards = _____ feet

5. 15 min = _____ hour

6. 5.5 miles = _____ ft

7. 9 meters = _____ centimeters

8. 424 milliliters = _____ liter

9. 0.5 kilogram = _____ grams

10. 0.01 liter = _____ milliters

11. Joe weighs 210 pounds. What is his weight in kilogram ?

12. Kathy is 5 feet 8 inches tall. What is her height in centimeters ?

13. Distance between Bangkok to Pattaya is about 145 miles. What is this distance in kilometers ?

Chapter 15
Probability

Probability is the chance of something happening.
Probability can be expressed as fraction, ratio, decimal or percentage.

$$\text{Probability of an event} = \frac{number\ of\ favorable\ outcomes}{number\ of\ possible\ outcomes}$$

Sum of probabilities = 1 or 100%

Example 1

What is the probability of tossing a coin and getting a head ?

number of possible outcome = 1 face head + 1 face tail = 2 possibilities

$$\text{Probability of head} = \frac{number\ of\ favorable\ outcomes}{number\ of\ possible\ outcomes} = \frac{1\ face\ head}{2\ possiblities} = \frac{1}{2} = 50\%$$

Example 2

Jack rolls a dice in a casino:

a) What is the probability that he gets a 2 from the roll ?

number of possible outcome = 6 possibilities (6 faces)

$$\text{Probability of 2} = \frac{number\ of\ favorable\ outcomes}{number\ of\ possible\ outcomes} = \frac{1\ face\ with\ 2}{6\ possiblities} = \frac{1}{6} = 16.7\%$$

b) What is the probability that he gets an odd number ?

$$\text{Probability of odd} = \frac{3\ face\ with\ odd}{6\ possiblities} = \frac{3}{6} = \frac{1}{2} = 50\%$$

Probability with *and* & *or*.

"And" means multiply (and → x)

"Or" means add (or → +)

Example 3

Maria has three pair of red socks, five pairs of green socks and two pairs of blue socks.

a) What is the probability that she picks a pair of red or a green socks?

$$P(Red \ or \ Green) \ = \ P(\ Red) + P(Green)$$

$$= \ 3/10 \ + \ 5/10$$

$$= \ 8/10$$

$$= \ 0.8$$

b) What is the probability that the first pair of socks she pick is red (without putting it back)and the second one is blue ?

$$P(Red \ _{1st} \ and \ Blue_{2nd} \ ) \ = \ P(Red^{1st}) \ x \ P(Blue^{2nd})$$

$$= \ 3/10 \ x \ 2/9$$

$$= \ 1/15$$

Example 4

John roll a dice and toss a coin at the same time. What is the probability that he gets a 4 on a dice and a tail on the coin ?

$$P(4 \ on \ a \ dice \ and \ tail \ on \ a \ coin) \ = \ P(4) \ x \ P(tail)$$

$$= \ 1/6 \ x \ 1/2$$

$$= \ 1/12$$

Exercise 15

1. When Adam throws a fair six sided dice:

    a) What is the probabilty of obtaining a 5?

    b) What is the probability of obtaining a 7?

    c) What is the probability of obtaining an even number?

    d) What is the probability of obtaining 2 or 4?

2. The spinner is divided into eight equal area.

    a) What is the probability of obtaining a 7?

    b) What is the probability of obtaining an odd number?

    c) What is the probability that it stops on a white area?

    d) What is the probability of obtaining 2 or 8?

3. Annabella has 4 red candies, 5 blue candies and 3 green candies. If all the candies are in her pocket, and she wishes to pick a candy:

    a) What is the probability that she picks a red or a green candies?

    b) What is the probability that she picks the first red and the second one green?

    c) What is the probability that she picks three blue candies without replacement?

4. Use the cards below to answer the questions

    a) What is the probability of picking an 'A'?

    b) If a queen card was removed then what is the probability of picking an 'A'?

    c) What is the probability of picking a king or a queen?

    d) What is the probability of picking a king and then a queen(without replacement)?

Revision 5

CHAPTER 13 to 15

1. A bag contains 6 nickles, 5 dimes, and 4 quarters. If one coin is drawn at random, what is the probability that the coin drawn is a dime?

a) 1/2
b) 1/3
c) 2/5
d) 4/15
e) 11/15

Information given below is for question 2 to 5:
The circle graph shows 7,200 wage earners in a small town during a given period.

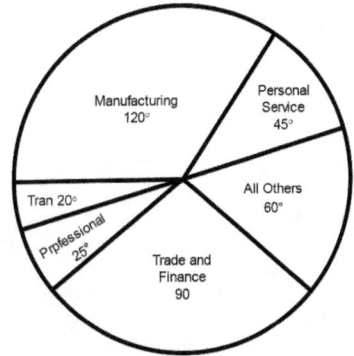

2. How many people earned their living in transportation field?

a) 400
b) 500
c) 900
d) 1200
e) 1500

3. How many more people worked in manufacturing than in personal service?

a) 400
b) 500
c) 900
d) 1200
e) 1500

4. What is the probability of randomly picking a person who worked in professional area?

a) 1/4
b) 1/6
c) 1/8
d) 2/3
e) 5/72

5. What is probability of randomly picking a person not in manufacturing field?

a) 1/4
b) 1/6
c) 1/8
d) 2/3
e) 5/6

6. On a certain high-way, five liter of oil was spilled by a truck. If a quarter of the spilled oil evaporated, then how many milliliters evaporated?

a) 250
b) 500
c) 1250
d) 2500
e) 3750

Information given below to answer question 7 to 9.

The graph shows income and expenses for each year indicated. The income is represented by shaded bar and the expenses by striped bars.

Income and Expense

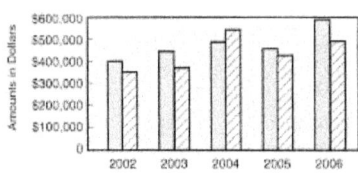

7. Which year did income exceeded the expenses by $100,000?

a) 2002
b) 2003
c) 2004
d) 2005
e) 2006

8. Which year did expense exceeded the income?

a) 2002
b) 2003
c) 2004
d) 2005
e) 2006

9. Which year was income and expense nearly equal to each other?

a) 2002
b) 2003
c) 2004
d) 2005
e) 2006

10. Mr.Smith buys a roll of cloth 20 feet 5 inches in length. He cuts the roll into 5 equal pieces, what is the length of each piece?

a) 4 feet
b) 4 feet 1 inch
c) 4 feet 5 inches
d) 7 feet
e) 7 feet 5 inches

11. What is the probability on rolling an eight sided dice number 1 to 8 and obtaining a prime number?

a) 1/8
b) 1/4
c) 1/3
d) 1/2
e) not enough information is given

12. To make a 7 inch pizza you need 2 pounds and 8 ounces of melted cheese. If Fred wanted to make four 7 inch pizza then how much cheese is needed?

a) 8 pounds
b) 8 pounds and 8 ounces
c) 10 pounds
d) 10 pounds and 8 ounces
e) 12 pounds

13. Light travel at the speed of 300,000 km per second, how many meter does it travel in 0.2 second?

a) 60,000
b) 80,000
c) 600,000
d) 800,000
e) 60,000,000

14. For every 2 days that it rains 3 days it will snow and 4 days it will be sunny. What is the probability that it will rain?

a) 1/2
b) 1/3
c) 2/3
d) 2/9
e) not enough information is given

15. A jar contain 4 red candies, 5 blue candies and 3 green candies. What is the probability that Joanna eats one blue and one green candy?

a) 8/12
b) 8/11
c) 5/22
d) 5/44
e) 7/55

16. Mark walks at an average speed of 5 miles per hour, at this rate how long will it take him to walk 13,200 feet?

a) 0.5 hour
b) 1.5 hours
c) 2 hours
d) 2.5 hours
e) 3.5 hours

17. How many second are there in one day?

a) 1,440
b) 32,400
c) 86,400
d) 172,800
e) 360,000

Graph given below is for question 18 to 20: The graph indicates the number of gallons of milk sold at a grocery store in 1 week period.

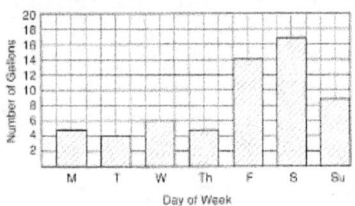

18. How many gallons of milk were sold on Thursday?

a) 4
b) 5
c) 6
d) 7
e) 14

19. How much more milk were sold on Friday than on Monday?

a) 4
b) 5
c) 9
d) 10
e) 14

20. How many gallons of milk were sold in one week?

a) 40
b) 55
c) 60
d) 75
e) 80

Chapter 16
Basic Geometry

<u>**Type of Lines**</u>

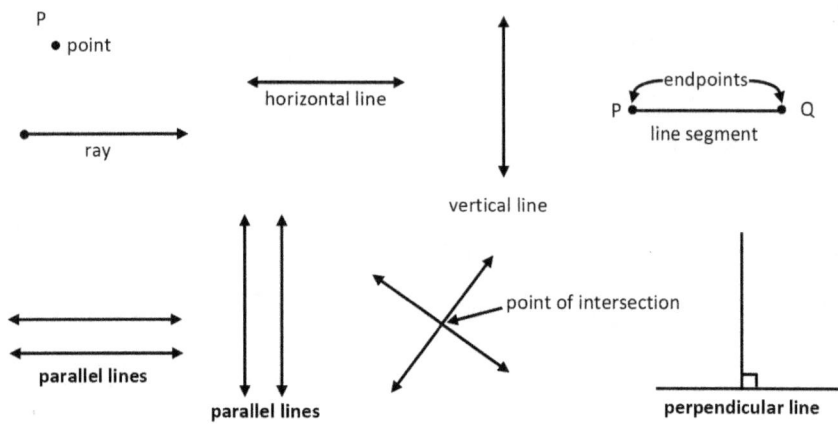

<u>**Type of angles**</u>

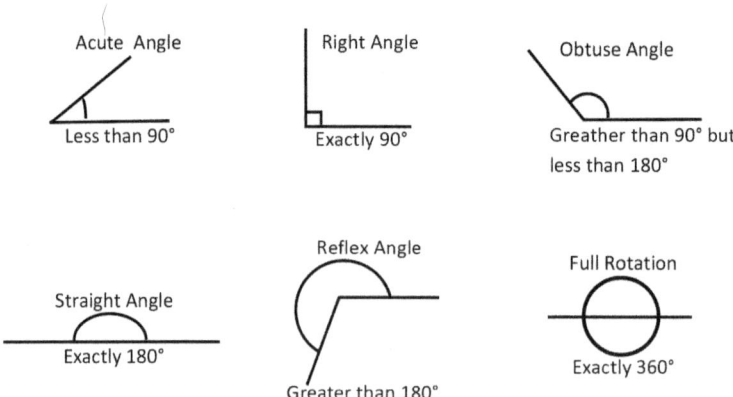

Triangles

Based on sides	
Scalene triangle All 3 sides have different lengths. Its angles are also all different	
Isosceles triangles 2 sides have equal lengths. 2 of its angles also measure equal.	
Equilateral triangle All 3 sides are of same length. All three angles are equal to 60°	
Based on angles	
Acute triangles All angles are less than 90°	
Right triangle Has 1 right angle (90°)	
Obtuse triangle Has one angle more that 90°	

Quadrilaterals

Quadrilateral	Properties	
Rectangle	4 right angles and opposite side equal	
Square	4 right angles and 4 equal sides	
Parallelogram	Two pairs of parallel sides and opposite sides equal	
Rhombus	Parallelogram with 4 equal sides	
Trapezium or Trapezoid	Two sides are parallel	
Kite	Two pairs of adjacent sides of the same length	

Circles

circumference

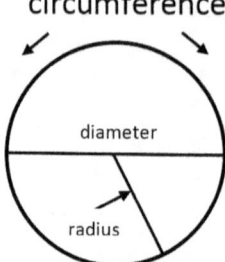

Rules of equivalent angles in parallel line

Rule 1: Alternate Interior Angles are equal

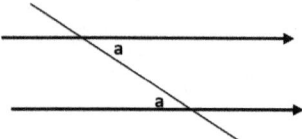

Rule 2: Vertically Opposite Angles are equal

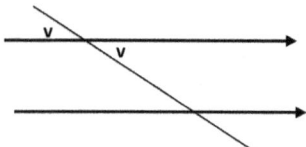

Rule 3: Alternate Exterior Angles are equal

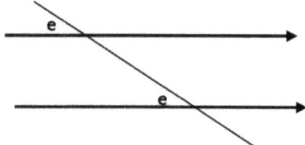

Exercise 16

Match the following letters to the statements.

a. 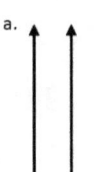 b. 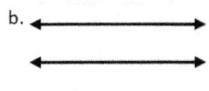 c.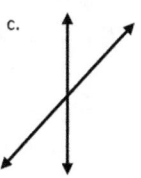

d. 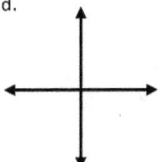 e.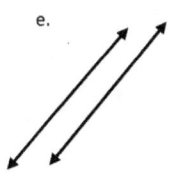

___1. intersecting but not perpendicular

___2. vertical and parallel

___3. parallel but neither vertical nor horizontal

___4. horizontal and parallel

___5. perpendicular

Identify each angle

6. a. 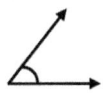 b. 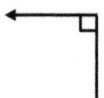 c.

7. a.    b. 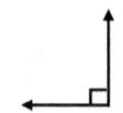   c.

8. a. 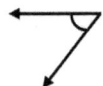   b.    c.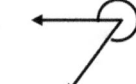

9. a. 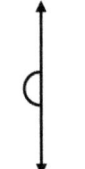   b. 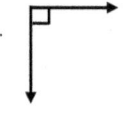   c.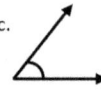

Identify each figure

10. 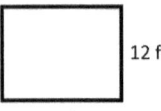 12 ft

15 ft

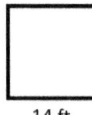

14 ft

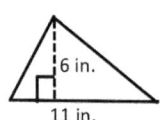 6 in.

11 in.

11.

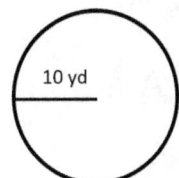

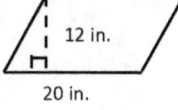

12.

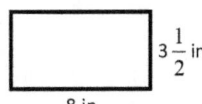

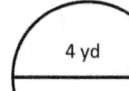

13. Find angles x, y and z

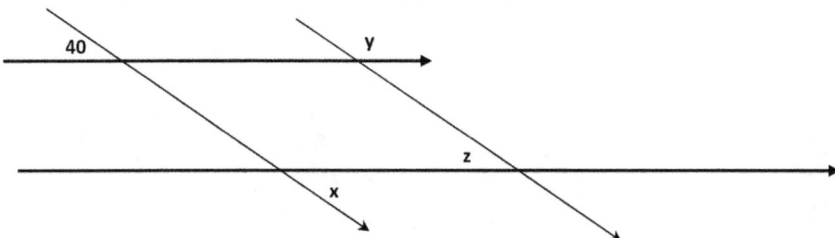

Chapter 17
Perimeter and Area

Triangle

Finding Area :

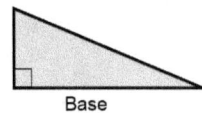

$$\text{Area} = \frac{1}{2} \times \text{Base} \times \text{Height}$$

Finding Perimeter :

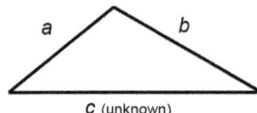

$$\text{Perimeter} = a + b + c$$

Types of triangle :

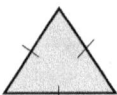

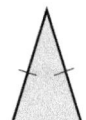

Equilateral triangle:

i) All sides are equal

ii) All angles are equal (each = 60)

Isosceles triangle:

i) Two sides are equal

ii) Two angles are equal

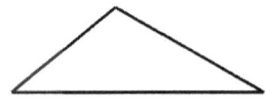

Scalene triangle:

i)No sides are equal

ii)No angles are equal

Circles

circumference

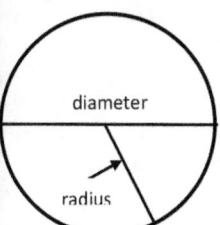

diameter

radius

Area = πr^2

Circumference = $2\pi r$

Quadilaterals

Finding Area:

		Formula
Square	Square s	s^2
Rectangle	Rectangle w	lw
Parallelogram	Parallelogram h	bh
Trapezoid	Trapezoid h	$\frac{1}{2}(a+b)h$

Sum of Interior Angle:

(n-2)x180 where "*n*" the number of sides or angles

Interior angle of a triangle is

(3-2)x180 = 1x180 = 180°

Interior angle of quadrilateral is

(4-2)x180 = 2x180 = 360°

Interior angle of pentagon is

(5-2)x180 = 3x180 = 540°

The Summary Table

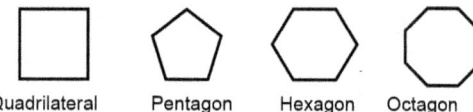

| | Quadrilateral | Pentagon | Hexagon | Octagon |

n	4	5	6	8
$(n-2)\times180$	$(4-2)\times180$	$(5-2)\times180$	$(6-2)\times180$	$(8-2)\times180$
Interior Angle	360°	540°	720°	1080°

Pythagoras Theorem

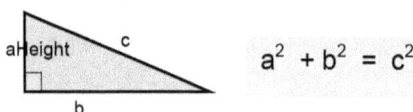

$$a^2 + b^2 = c^2$$

Exercise 17

Find the area of each figure

1. 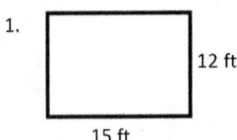 12 ft

   15 ft

   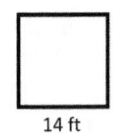

   14 ft

   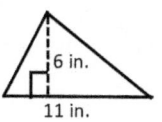 6 in.

   11 in.

2. 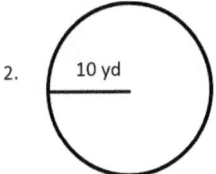 10 yd

   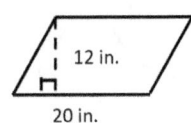 12 in.

   20 in.

   14 in.

   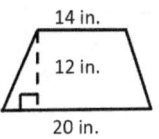 12 in.

   20 in.

3. 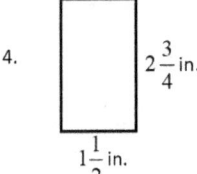 $3\frac{1}{2}$ in.

   8 in.

   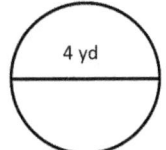 4 yd

   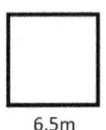

   6.5m

Find the perimeter of each figure

4. 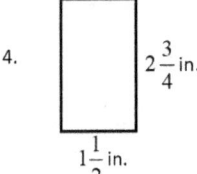 $2\frac{3}{4}$ in.

   $1\frac{1}{2}$ in.

   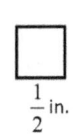

   $\frac{1}{2}$ in.

   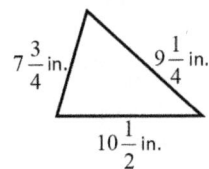 $7\frac{3}{4}$ in. $9\frac{1}{4}$ in.

   $10\frac{1}{2}$ in.

5. 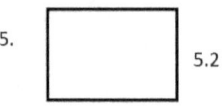 5.2 cm

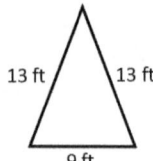 1.75 m

13 ft 13 ft

9 ft

6.3 cm

6. 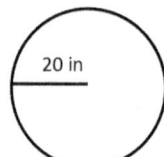 10yd

20 in

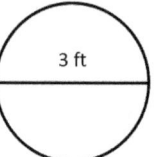 3 ft

7. Find the perimeter, in inches, of the figure shown at the right.

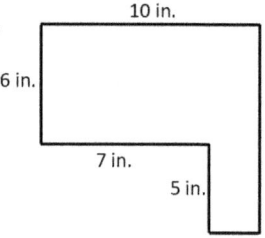

10 in.

6 in.

7 in.

5 in.

8. How many square inches are on the surface of the figure in the last example?

9. Find the area of the figure shown at the right.

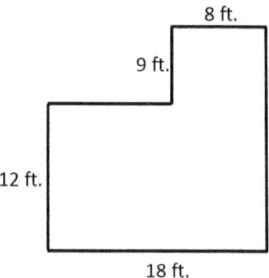

8 ft.

9 ft.

12 ft.

18 ft.

10. What is the minimum number of 4-inch by 4-inch square tiles that are needed to cover the cover the top of a square coffee table that measures 5 feet on each side?

11. Find the value of x

a)

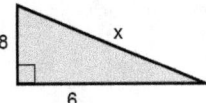

b)

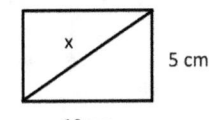

Revision 6

CHAPTER 16 to 17

1. In triangle ABC What is the measure of angle C

a) 50
b) 60
c) 70
d) 80
e) 90

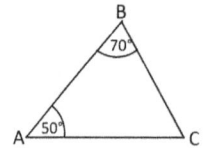

2. Which expression gives the perimeter of the figure below?

a) $6(2) + \pi (6)$
b) $6(3) + \pi (3)$
c) $6(3) + \pi (6)$
d) $6(4) + \pi (3)$
e) $6(4) + \pi (6)$

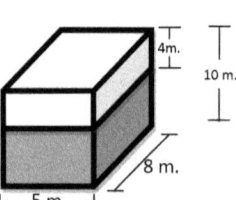

3. What is the volume of the water in the tank?

a) 160
b) 200
c) 240
d) 360
e) 400

4. What is the sum of the interior angle of nonagon (9-side polygon)?

a) 720°
b) 900°
c) 1080°
d) 1260°
e) 1440°

5. If △ABC is similar to ∠YZ, what is the length of YZ?

a) 12
b) 15
c) 22
d) 30
e) 33

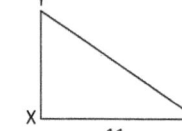

6. Find the length of the diagonal of a 6 by 8 rectangle?

a) 5
b) 10
c) 14
d) 28
e) 48

7. If AB is parallel to CD, what is the size of angle CAB?

a) 110°
b) 100°
c) 90°
d) 80°
e) 70°

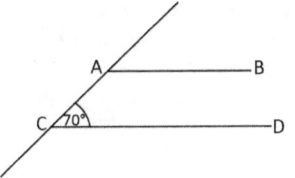

8. What is the perimeter of figure below?

a) 22x
b) 25x
c) 29x
d) 32x
e) 37x

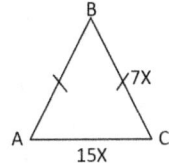

9. What is the volume of the wooden incline?

a) 540
b) 600
c) 720
d) 900
e) 1800

10. What is the diameter of the circle with the area of 48 π?

a) $\sqrt{3}$
b) $2\sqrt{3}$
c) $3\sqrt{3}$
d) $4\sqrt{3}$
e) $8\sqrt{3}$

11. How many 4 inch by 5 inch by 1 inch CD can be kept in a box that is 4 ft. by 5 ft. by 1 inch?

a) 144
b) 64
c) 50
d) 20
e) 10

12. Which expression shows the correct relationship between x, y and z?

a) x − y − z = 180
b) x + y + 180 = z
c) 360 − x − y − z = 180
d) x + y + z + 180 = 0
e) x = -y + z − 180

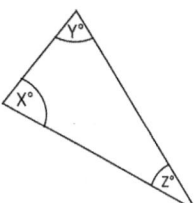

13. A square vegetable farm has an area of 169 square foot. If a fence is to be put around the farm, how long should the fence be?

a) 13
b) 26
c) 39
d) 52
e) 65

14. For diagram below, ∠ BAC = 80° and

∠ EAC = 30°. If line segment AD bisects ∠ BAC, what is the measure of ∠ DAE?

a) 10°
b) 20°
c) 30°
d) 40°
e) 50°

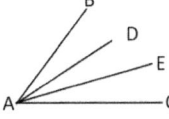

15. Jacob drives 50 km East, then 120 km North. How many kilometer are saved if he drives directly to the destination?

a) 20 km.
b) 40 km.
c) 50 km.
d) 70 km.
e) 130 km.

16. Find the area of the triangle to the nearest unit.

a) 52
b) 49
c) 26
d) 20
e) 10

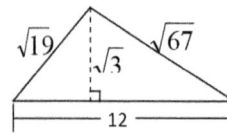

17. Find the area of triangle ABC.

a) 18
b) 90
c) 320
d) 540
e) 960

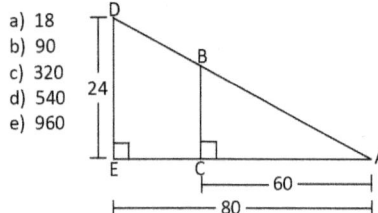

18. What is the radius of the large circle, if the small circles are identical and tangent to each other and has on area of 9π each?

a) 3
b) 6
c) 9
d) 12
e) 18

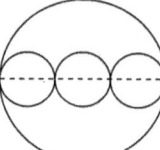

19. If line p and q are parallel. What is the value of x?

a) 20
b) 30
c) 40
d) 50
e) 60

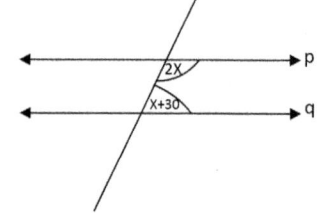

20. What is the volume of a hemisphere with radius 6 cm.?

a) 36π
b) 108π
c) 144π
d) 288π
e) 864π

Chapter 18
3-D Geometry

Volume:

Volume = surface area x depth

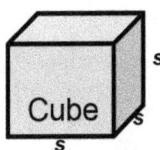

Volume = s^3

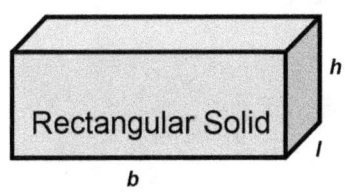

Volume = $b \times l \times h$

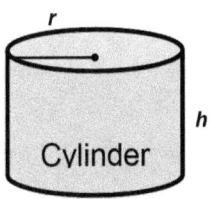

Volume = $\pi r^2 h$

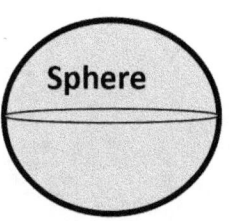

Volume = 4/3 πr^3

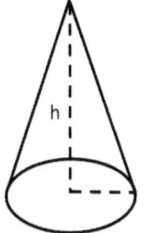

Volume = 1/3(base area)h

Volume of a :

Cube	Volume = edge^3
Rectangular solid	Volume = length x width x height
Square pyramid	Volume = $\frac{1}{3}$ x (base edge)^2 x height
Cylinder	Volume = π x radius^2 x height; $\pi \approx 3.14$
Cone	Volume = $\frac{1}{3}$x π x radius^2 x height; $\pi \approx 3.14$

Exercise 18

Find volume of each figure

1.

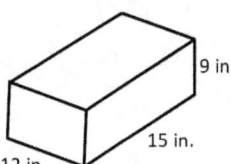

9 in.

15 in.

12 in.

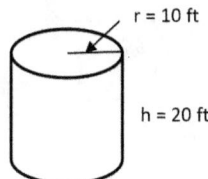

8 in.

8 in.   8 in.

r = 10 ft

h = 20 ft

2.

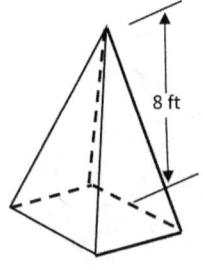

8 ft

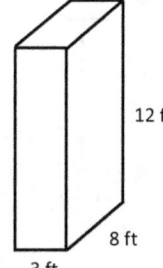

12 ft

8 ft

3 ft

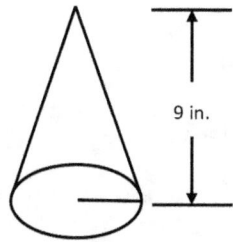

9 in.

3.

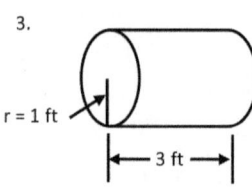

r = 1 ft

|← 3 ft →|

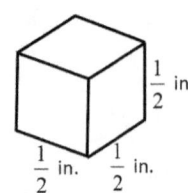

$\frac{1}{2}$ in.

$\frac{1}{2}$ in. $\frac{1}{2}$ in.

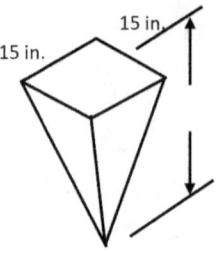

15 in.

15 in.

4.

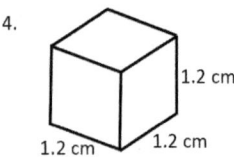

1.2 cm

1.2 cm   1.2 cm

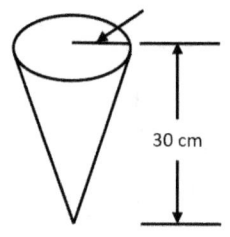

30 cm

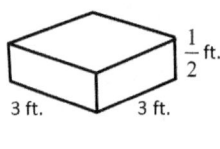

$\frac{1}{2}$ ft.

3 ft.   3 ft.

5. What is the volume, in cubic inches, of the carton shown below?

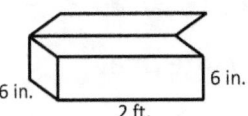

6 in. 6 in.
2 ft.

6. How many boxes like the one below can fit into the carton in the last problem?

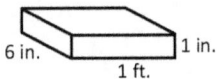

6 in. 1 in.
1 ft.

7. The diagram at the right shows a large rectangular block with a cylindrical hole through it. Which of following expresses the volume, in cubic feet, of the concrete that is required to construct the block?

(1) $15 \times 10 \times 10 - \pi \times 3^2 \times 15$

(2) $10 \times 10 - \pi \times 3^2$

(3) $15 \times 10 - \pi \times 3^2 \times 15$

(4) $15 \times 10 - 3^2 \times 15$

(5) $\pi \times 3^2 \times 15$

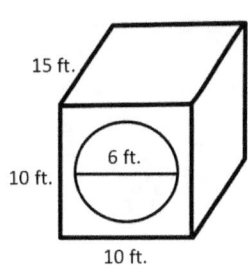

15 ft.

10 ft.

6 ft.

10 ft.

Chapter 19

Basic Algebra

Example 1

Simplify $5x + 6 - 2x - 10 + 2y$

Step1:   Group the like terms

$$= 5x - 2x + 2y + 6 - 10$$

Step2: Do the operation

$$= 3x + 2y - 4$$

Answer: $3x + 2y - 4$

Example 2

Solve the following $2x - 3 = 12 - 3x$

Step1: Separate the numbers and variables to one side

$$2x + 3x = 12 + 3$$

Step2: Do the operation

$$5x = 15$$

Answer: $x = 3$

Example 3

If $2 - 3x < 5$, find all possible values of x?

$$2 - 3x < 5$$

$$- 3x < 5 - 2$$

$$- 3x < 3$$

$$x > -1 \quad \text{(swap inequality sign when multiply or divide by a negative number)}$$

Answer: $x > -1$

Word translation into algebra:

A number increased by four	$x + 4$
Nine more than a number	$x + 9$
Eight less than a number	$x - 8$
Eight decreased by a number	$8 - x$
Four times a number	$4x$
The product of ten and a number	$10x$
A number divided by five	$\dfrac{x}{5}$
One-half of a number	$\dfrac{x}{2}$
One-fourth of a number	$\dfrac{x}{4}$

Example 4

Two less than three times a number is equal to seventy, find twice of that number.

Step1: Translate English to Math equations

$$3x - 2 = 70$$

Step2: Solve for x

$$3x = 70 + 2$$

$$x = 72/3$$

$$x = 24$$

Step3: Read the question (Find twice a number)

$$2x = ???$$

$$2(24) = 48$$

Answer: 48

Exercise 19

1. Simplify the following :

   a) $12x - 7y + 3(x - 5y + 2)$

   b) $4(a - 2c) - 3(c - 3a)$

   c) $4r^2 + 5r - 2(3r^2 - 4.5r)$

2. Solve for each unknown

   a) $12.5 = 2y + 4.5$

   b) $96 + 2x = x + 101$

   c) $3(a - 4) = 5a + 11$

3. Solve each equation/inequality

   a) $9a - 5 > 7a + 11$

   b) $5/x = x/20$

   c) $4r^2 - 25 = 0$

4. Translate to algebra and solve the following

   a) Nine times a number decreased by five equal six times the same number increased by seven, what is the number ?

   b) Ten times a number decreased by seven equals 101 plus that number. What is the number?

   c) Doubling a number result in a number increased by the product of seven and nine. What is the number ?

5. Jack and Jill are programmers. Jack makes $ 5 an hour less than Jill. On a job that took them both 40 hours to do, they made $ 1,400. How much does Jack make in one hour?

6. Leo makes $ 380 more each month than his wife. Together they make $ 3200 each month. How much does Leo make in a month?

Chapter 20
Linear and Quadratic

The Graph of Linear Function:

Linear means graph of a straight line.

Linear Equation:

$y = mx + c$

"c" is a y-intercept, where the graph crosses y-axis

"m" is a gradient or slope

$$\text{Slope} = \frac{\Delta Y}{\Delta X} = \frac{y_2 - y_1}{x_2 - x_1}$$

 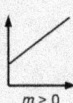 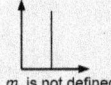

$m = 0$ $m > 0$ $m < 0$ m is not defined

Example 1

Identify the slope, y-intercept and plot the graph of: y = 3x - 4

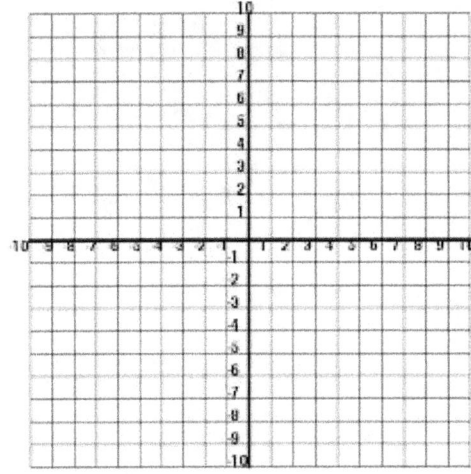

Example2

Find the equation of the line that passes through point (2,5) and (6,9)

Then plot the graph and locate its midpoint

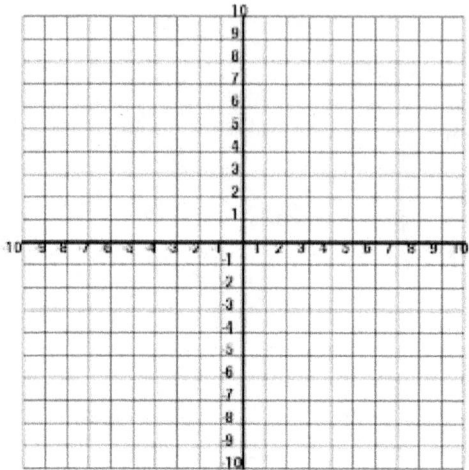

Quadratic equation:

$$(x + y)^2 = x^2 + 2xy + y^2$$

$$(x - y)^2 = x^2 - 2xy + y^2$$

$$(x + y)(x - y) = x^2 - y^2$$

Example3

If $x^2 - 2x = 3$, what is a positive value of x?

Solution:

$$x^2 - 2x = 3$$

$(x + a)(x + b) = x^2 + (a + b)x + (ab)$

$$x^2 - 2x - 3 = 0$$

$$(x - 3)(x + 1) = 0$$

$$x = 3 \text{ or } -1$$

Therefore, a positive value of x is 3.

Example 4

Graph equation $y = x^2 - 2x - 3$

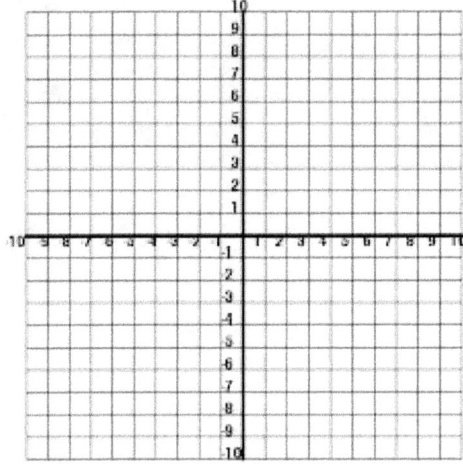

Exercise 20

1. Find the slope of:

    a) (-3, 2), (5, -10)

    b) (5,4) , (0,4)

    c) (u, 3u) , (-2u, 4u)

2. Find the equation of the line that has:

    a) slope of 3.5 and y-intercept of -2

    b) pass through (-3, 2) and (5, -10)

    c) x-intercept of -4 and no y-intercept

3. Plot the graph of

    a) $2y = -3x + 4$ b) $5x = 20y$

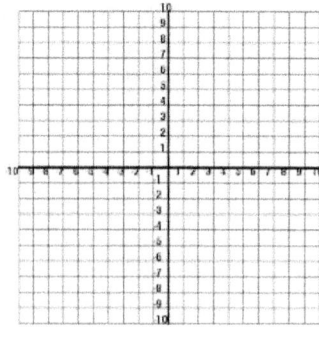 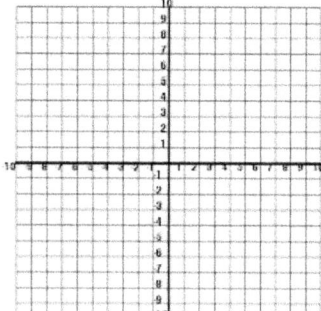

4. Solve the following

    a) $x^2 - 5x = 6$

    b) $3x^2 + 11x - 4 = 0$

5. Graph the following

a) $y = x^2 - 2x$

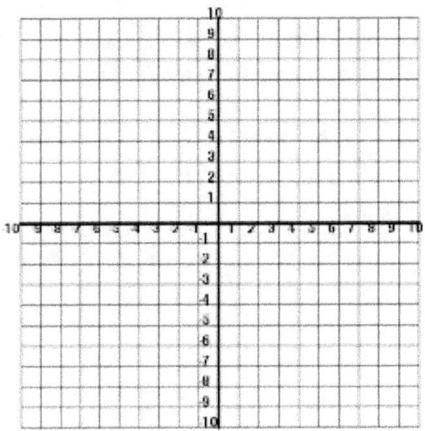

b) $y = 2x^2 - 5$

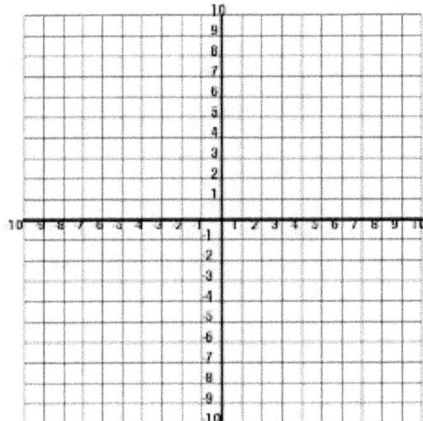

1. Let x, y, and z be a different number, which of the following equations has the some result as "x minus 3z equals y"

a) x - 3y = z
b) x + 3y = z
c) x − z = y
d) x + 3z = y
e) x = y + 3z

2. There are shirts of 5 colors and each with 4 sizes, how many different color and size combination are possible for the shirts?

a) 5
b) 9
c) 20
d) 26
e) 35

3. What is the slope of the line that passes thought point (b,a) and (c,a)?

a) $\frac{c-b}{2a}$
b) $\frac{b-c}{a}$
c) $\frac{2a}{c-b}$
d) 0
e) undefined

4. If 5a − 3 < 12, then how many positive integer value does "a" have?

a) 1
b) 2
c) 3
d) 4
e) 5

5. Let P(x) be the profit function of producing x units of goods. Given $P(x) = 5x^2 - 20$. Find the profit when 20 units are being produced?

a) 80
b) 120
c) 240
d) 1280
e) 1980

6. On the grid below locate the y-intercept of graph of 7y = 7x - 21?

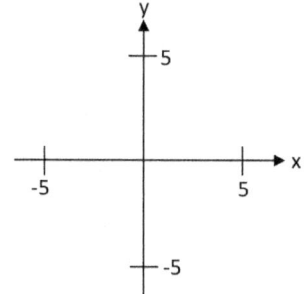

7. Which of the following represents "twice of a number increased by twelve is equal half number"?

a) 2(N + 12) = 0.5N
b) 2N + 12 = 0.5N
c) 2N − 12 = 5N
d) 2N + 2 = 0.5N
e) 2N + 5 = 12N

8. What is the equation of the line that contains point (2, -3) and (0, 2)?

a) 5x + 2y = 4
b) 3x + 2y = 6
c) x + 2y = 8
d) 5x − 2y = -4
e) 3x − 2y = -6

9. Solve for x, $5x^2 − 125 = 0$?

a) x = 0, x = 5
b) x = 0, x = 25
c) x = 5, x = 25
d) x = -5, x = 5
e) x = -5, x = 0

10. What is the equation of the line that has a y-intercept at 5 and pass through (-5, 5)?

a) y = 5x + 5
b) y = 5x
c) y = x + 5
d) x = 5
e) y = 5

11. Draw the graph of $y = x^2 − 4$ on the grid below

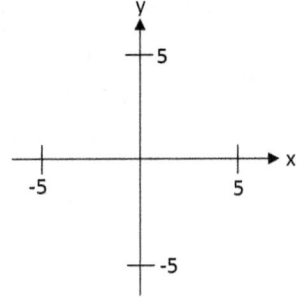

12. If a = 2, b = 3, c =-1 What is the value of $3a^2 − b + 5c$?

a) 3
b) 4
c) 8
d) 10
e) 20

13. If x is a negative number, which choice will make it a positive number?

a) x^3
b) 5x
c) $\frac{x}{3}$
d) x + 2
e) 3 − x

14. If x is a positive integer, which of the following gives the smallest value?

a) $\frac{x}{01}$
b) $\frac{x}{001}$
c) $\frac{x}{10}$
d) $\frac{x}{100}$
e) $\frac{x}{1000}$

15. If 2w−0.04 = 1.24, what is the value of w?

a) 2.56
b) 2.40
c) 1.28
d) 0.64
e) 0.60

16. If $2^a = 8^2$, what is the value of a?

a) 3
b) 4
c) 5
d) 6
e) 8

17. Point B is at (3, -2) and is reflected on y-axis, what is the new coordinate of point B?

a) (-3, -2)
b) (-3, 2)
c) (3, 2)
d) (2, 3)
e) (2, -3)

18. What value of x will make the equation $|x - 3| = 10$ true?

a) 13
b) 7
c) 4
d) -13
e) -6

19. What is the equation of the line parallel to line $y = \frac{1}{3}x - 4$ that has a y-intercept of 2?

a) $y = \frac{1}{3}x - 2$
b) $y = \frac{1}{3}x + 2$
c) $y = 3x - 2$
d) $y = 3x + 2$
e) $y = -3x + 2$

20. If $a^2 - a - 6 = 0$ what is a positive root of "a" equal to?

a) 0
b) 1
c) 2
d) 3
e) 4

Post-Test

PART 1 (calculator can be used)
Question 1 through 3 refer to the following information.
Annie's Ice-Cream has 4 stores located in different sections of the town. The numbers of ice-cream sold at each location for 5 days are shown in the table below.
Number of ice-cream sold

Store	Mon	Tue	Wed	Th	Fri
A	78	101	98	105	159
B	85	111	96	102	119
C	90	99	89	111	165
D	67	103	100	97	145

1. Annie is thinking of advertising a sale promotion on the radio. For which day should this be done?

a) Monday
b) Tuesday
c) Wednesday
d) Thursday
e) Friday

2. What was the mean number of ice-cream sold on Friday?

a) 145
b) 147
c) 153
d) 155
e) 160

3. How much more ice-cream were sold on Friday than on Tuesday?

a) 134
b) 144
c) 154
d) 164
e) 174

4. A pendulum is moving with a 5 meter long string. Approximately how far, in meters, does the bob swing?

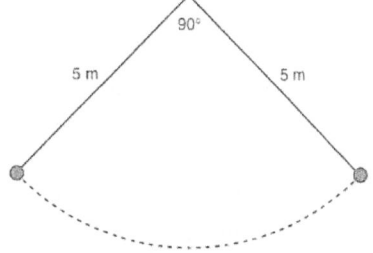

a) 5.4
b) 7.9
c) 8.2
d) 14.5
e) 31.4

5. Evaluate 25 + 5 ÷ 0.1 x 2 -3?

a) 97
b) 122
c) 165
d) 349
e) 597

6. Juan's gross weekly salary is $ 300, but 20% is withheld for taxes and provident funds. She sets aside $ 65 from her take-home pay each week for her parents. Juan budgets one-fifth of the remainder for miscellaneous expenses, and she puts the rest in her saving account.

How much per week does Juan save?

a) 85
b) 140
c) 175
d) 195
e) 240

7. The area of triangle WXZ in the diagram below is 80 centimeter square

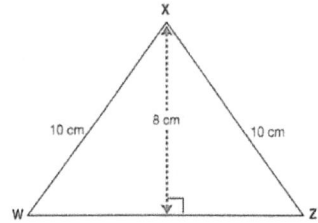

What is the perimeter of the triangle?

a) 20
b) 30
c) 40
d) 48
e) 58

8. Three cans of coke and a can of beer cost total of $ 5.00. Two cans of coke and three cans of beer cost $8.00.   What is the cost of a can of beer?

a) 1
b) 1.5
c) 2
d) 2.5
e) 3

9. A coffee shop menu lists 5 type of coffees, 4 type of appetizers, and 3 type of spaghettis. Each day the coffee shop offers a combination of one of each at a special price. How many combination are possible when one of each kind is taken?

a) 12
b) 18
c) 36
d) 60
e) 80

10. Natty's cellular phone service has a monthly charge of $ 40.99 but adds a service charge of $ 0.25 per minute of calls. Natty has budgeted $ 55 per month for using her cell phone.

What is the maximum number of minutes per month that she can call?

a) 11
b) 14
c) 31
d) 56
e) 72

PART 2 (no calculator)

11. Evaluate $32 \times 10^{-2} \times 0.5 \times 10^3$?

a) 0.16
b) 1.6
c) 16
d) 160
e) 1600

12. Two rectangle have equal areas. The first rectangle has a length of 1.2 feet and a width of 5 feet. The length of the second rectangle is 15 inch. What is the width, in feet, of the second rectangle?

a) 1.6
b) 2.5
c) 3.2
d) 4.8
e) not enough information is given

13. Nanny is dissatisfied with her current monthly salary. If she received a raise of $5 per hour and worked an additional 20 hours each week, her monthly salary would then be $600. What is Nanny's current weekly salary?

a) 300
b) 400
c) 500
d) 550
e) not enough information is given

Information given below is for question 14 to 15:

Type of gold	Price per oz.	Price per 10 oz.
16k	$ 32.5	$ 299
22k	$ 43.25	$ 423.50

14. Lara wants to buy 11 oz. of 16k and 12 oz. of 22k, what is the cheapest price she must pay?

a) 714.43
b) 818.71
c) 822.02
d) 841.50
e) 904.22

15. If Lara wants to make a statue of 22k of gold and she has a maximum of $1730 to spend, how many ounces of gold will the statue use?

a) 30
b) 32
c) 38
d) 40
e) 51

16. In square ABCD below points X and Y are midpoints of each side?

What is the ratio of the shaded area to the

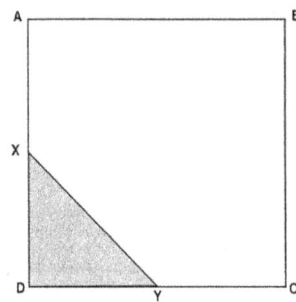

white area?

a) 1 : 4
b) 1 : 5
c) 1 : 7
d) 2 : 5
e) 2 : 7

17. Water is pumped into an empty tank at a constant rate. After 2 hours, the tank contains 300 gallons. After 5 hours, the tank is full. How many gallons of water does the full tank hold?

a) 600
b) 650
c) 700
d) 750
e) 800

18. What is the sum of the third and fourth number in the box below?

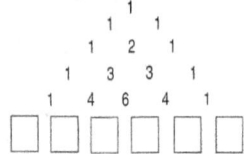

a) 10
b) 18
c) 20
d) 24
e) 32

19. Which of the following is the complete solution $(x - 12)(x + 5) = 0$?

a) - 5 only
b) 12 only
c) 5 and -12
d) -5 and 12
e) -5 and -12

20. Luke wanted to post ads on local newspaper. The news company charges $2 for the first 45 words and ten cent per word thereafter. How much does Luke pay if he advertises using 93 words?

a) $ 4
b) $ 4.30
c) $ 4.80
d) $ 6
e) $ 6.80

Formula :

<u>Arithmetic</u>

Mean: $\qquad$ Mean $= \dfrac{X1+X2+X3+\ldots+Xn}{n}$

Percentage change: $\%Change = \dfrac{(New-Original)\times 100\%}{Original}$

Simple interest: $Interest = \dfrac{(Principal\ x\ Rate\ x\ Time)}{100}$

Probability: $\quad$ Probability of an event $= \dfrac{number\ of\ favorable\ outcomes}{number\ of\ possible\ outcomes}$

<u>Geometry (2D)</u>

Area:

Area of triangle $= \dfrac{1}{2} \times Base \times Height$

Area of square $= Side^2$

Area of rectangle $= Length\ x\ Width$

Area of parallelogram $= Base\ x\ Height$

Area of trapezium $= \dfrac{1}{2} \times (Base1 + Base2) \times Height$

Sum of Interior Angle:

$(n-2)x180$

where "n" the number of sides or angles

Geometry (3D)

Volume :

Volume of Cube = edge^3

Volume of Rectangular solid = length x width x height

Volume of Square pyramid = $\frac{1}{3}$ x (base edge)^2 x height

Volume of Cylinder = π x radius^2 x height

Volume of Cone = $\frac{1}{3}$x π x radius^2 x height

Algebra

Linear Equation:

$y = mx + c$

"*c*" is a y-intercept, where the graph crosses y-axis

"*m*" is a gradient or slope

$$\text{Slope} = \frac{\Delta Y}{\Delta X} = \frac{y_2 - y_1}{x_2 - x_1}$$

Quadratic Equation:

$(x + y)^2 = x^2 + 2xy + y^2$

$(x - y)^2 = x^2 - 2xy + y^2$

$(x + y)(x - y) = x^2 - y^2$

Keywords:

A

absolute value - the distance from a number listed to zero on the number line

acute angle - an angle that has a measure of less than 90 degree

area - measure of the amount of surface on a closed plan figure

average (mean) - the sum of a set of numbers divided by the number of numbers

C

circumference - the distance around a circle

common denominator - the bottom part of fractions that must be of same multiple value

complementary angles - angles whose sum is equal to 90 degree

congruent figures - geometric figures of the same shape, same size and same angle

consecutive - next to each other in ascending way

coordinate plane - a surface divided by x-axis and y-axis

coordinates - a pair of numbers in the form of (x,y)

D

degree - unit of measurement of angles

denominator - the bottom part of a fraction

diameter - the distance across and through the center of a circle (or twice of radius)

E

equilateral triangle - a triangle whose side are all equal and each angle measures 60 degree

estimate - an approximate value

even number - a number that is divisible by 2

exponent - a power of a number

F

factor - the product(s) that the number is made of

H

hypotenuse - the side of a right triangle that is opposite to the right angle and is the longest side

I

improper fraction - a fraction which the numerator and denominator can still be reduced to lowest form or mixed number

integer - number that contain no decimal point

intercept - the coordinates of point where a line crosses x-axis or y-axis

isosceles triangle - a triangle whose two sides are equal and the angle opposite to the two sides are also equal

L

legs - the two shorter sides of a right triangle

line segment - a line that has a starting point and an ending point

linear equation - an equation of a straight line

M

mean - the average value

median - the middle number(value) of a set of numbers

mode - a number that occurs the most in a set of numbers

N

numerator - the top number of the fraction

O

obtuse angle - an angle that measures more than 90 degree but less than 180 degree

odd number - a number that cannot be divided by 2

P

parabola - a curve graph of a quadratic equation

parallel lines - lines that run in the same-direction and do not cross each other

perimeter - the distance around a closed figure

perpendicular lines - lines that intersect each other at a right angle

pi - a Greek letter representing the ratio of circumference of a circle to its diameter, has an approximate value of 3.14 or 22/7

plane - a flat surface

polygon - a closed plane figure formed by three or more line segments meeting each other at their endpoints

prime number - number that has only two factors (itself and 1)

product - numbers that multiply each other

probability - the chance of an event happening, express in ratio or percentage

proportion - two fractions or numbers in ratio of each other

protractor - a tool used for measuring and drawing angle in degree

Q

quadratic equation - equation of a parabola that has one of the variable raised to power of 2

quotient - a division of numbers

R

radius - a distance from the center of a circle to its circumference

ratio - comparison of two numbers (division of two numbers)

reflex angle - angle that measures more than 180 degree but less than 360 degree

right angle - an angle that measures exactly 90 degree

right triangle - a triangle with one of the angle equal to 90 degree

rounding - estimating a number close to the original value

S

scalene triangle - a triangle with no equal sides and no equal angles

scientific notation - a number written as a product of a number between 1 to 10 with a product of 10 to a certain integer power

similar figure - geometric figures with same shape but not equal size

slope - a measure of steepness or rise/run

straight angle - an angle that measures exactly 180 degree

sum - addition value

supplementary angle - angles whose sum is 180 degree

T

transversal - a line that cuts across parallel lines

trigonometry - a study of relationship between sides and angles of a triangle

V

variable - a symbol or a letter that represents a number

vertex - the point where two sides of line segment meet

volume - amount of space inside 3-D figure

W

whole number - zero and a positive integer

Answer Key

Revision 1:
1. b
2. d
3. b
4. c
5. d
6. a
7. a
8. b
9. b
10. c
11. d
12. e
13. d
14. e
15. b
16. e
17. a
18. d
19. d
20. c

Revision 2:
1. b
2. e
3. c
4. c
5. c
6. e
7. d
8. b
9. a
10. c
11. c
12. d
13. a
14. d
15. d
16. c
17. e
18. c
19. e
20. d

Revision 3:
1. c
2. e
3. e
4. a
5. d
6. b
7. d
8. b
9. e
10. d
11. b
12. c
13. a
14. b
15. b
16. d
17. a
18. a
19. b
20. b

Revision 4:
1. e
2. b
3. d
4. b
5. a
6. d
7. d
8. b
9. a
10. a
11. c
12. c
13. d
14. c
15. e
16. d
17. b
18. e
19. c
20. b

Revision 5:
1. b
2. a
3. e
4. e
5. d
6. c
7. e
8. c
9. d
10. b
11. d
12. c
13. e
14. d
15. d
16. a
17. c
18. b
19. c
20. c

Revision 6:
1. b
2. b
3. c
4. d
5. e
6. b
7. a
8. c
9. d
10. e
11. a
12. c
13. d
14. a
15. b
16. e
17. d
18. c
19. d
20. c

Revision 7:
1. e
2. c
3. d
4. b
5. e
6. intercept at -3 on y-axis
7. b
8. a
9. d
10. e
11. parabola graph drawn
12. b
13. e
14. e
15. d
16. d
17. a
18. a
19. b
20. d

Post-Test:
1. a
2. b
3. e
4. b
5. b
6. b
7. c
8. c
9. d
10. d
11. d
12. d
13. e
14. d
15. d
16. c
17. c
18. c
19. d
20. e

www.ingramcontent.com/pod-product-compliance
Lightning Source LLC
Chambersburg PA
CBHW070042210526
45170CB00012B/567